INTRODUCTION TO G
AND TECHNOLOGY

RSC Paperbacks

RSC Paperbacks are a series of inexpensive texts suitable for teachers and students and give a clear, readable introduction to selected topics in chemistry. They should also appeal to the general chemist. For further information on selected titles contact:

Sales and Promotion Department
The Royal Society of Chemistry
Thomas Graham House
Science Park, Milton Road
Cambridge CB4 4WF

Titles Available

Water *by Felix Franks*
Analysis – What Analytical Chemists Do *by Julian Tyson*
Basic Principles of Colloid Science *by D. H. Everett*
Food – The Chemistry of Its Components (Third Edition)
by T. P. Coultate
The Chemistry of Polymers *by J. W. Nicholson*
Vitamin C – Its Chemistry and Biochemistry
by M. B. Davies, J. Austin, and D. A. Partridge
The Chemistry and Physics of Coatings
edited by A. R. Marrion
Ion Exchange: Theory and Practice (Second Edition)
by C. E. Harland
Trace Element Medicine and Chelation Therapy
by David M. Taylor and David R. Williams
Archaeological Chemistry
by A. M. Pollard and C. Heron
Introduction to Glass Science and Technology
by James E. Shelby

How to Obtain RSC Paperbacks

Existing titles may be obtained from the address below. Future titles may be obtained immediately on publication by placing a standing order for RSC Paperbacks. All orders should be addressed to:

The Royal Society of Chemistry
Turpin Distribution Services Ltd.
Blackhorse Road
Letchworth
Herts. SG6 1HN
Telephone: +44 (0) 1462 672555
Fax: +44 (0) 1462 480947

RSC Paperbacks

INTRODUCTION TO GLASS SCIENCE AND TECHNOLOGY

JAMES E. SHELBY

New York State College of Ceramics at Alfred University
2 Pine Street
Alfred, NY 14802, USA

THE ROYAL SOCIETY OF CHEMISTRY
Information Services

Dedication

To Dr. Delbert E. Day, my advisor and mentor, who taught me the wonders of glass. Thank you.

ISBN 0-85404-533-3

A catalogue record for this book is available from the British Library

Published by The Royal Society of Chemistry, Thomas Graham House, Science Park, Milton Road, Cambridge CB4 4WF, UK

Typeset by Computape (Pickering) Ltd, Pickering, North Yorkshire, UK
Printed and bound by Athenaeum Press Ltd, Gateshead, Tyne and Wear, UK

Preface

This book is intended as an introductory level text for the student or professional scientist or engineer interested in glass science and technology. It is assumed that the reader has little, if any, prior knowledge of glass science. As a direct consequence, the material is deliberately limited to that which can be covered in a single semester of course work. Restriction of the information to this level prevents the common problem of overwhelming the student by presenting material better left to an advanced level course.

The material in this text is presented in the order used when teaching 'Introduction to Glass Science' at the New York State College of Ceramics. In teaching this material, I find it useful to first define 'glass', specify those aspects of glasses which make them different from other materials, and discuss the historical development of glass technology. Following this introduction, the text addresses the questions of how glasses are actually produced and why some materials form glasses while others do not. The next two chapters of this text deal with the atomic arrangement and microstructure of glasses, with an emphasis on understanding of the basic principles of network structures and the details of phase separation. The next five chapters discuss the properties of glasses, including viscosity, thermal expansion and density, properties controlled by transport phenomena, mechanical properties, and optical behavior. The final chapter presents an overview of classical and specialized forming methods used to produce commercial products.

This text includes routine discussion of the effects of phase separation and crystallization on the properties of glasses. These effects are essentially neglected in all other texts on glasses. Since many modern glasses are phase separated, this neglect is no longer justified. Many papers published today have incorrectly interpreted the results of both spectral and property studies due to lack of understanding of the effects of phase separation on the details of the behavior of glasses.

The astute instructor will quickly note that this text does not include any problem sets. Problem sets were deliberately omitted in order to maximize use of the allotted space. Sets of problems for this text can be obtained by direct request to the author at the New York State College of Ceramics, Alfred, NY 14802, USA.

Contents

Acknowledgments

A large number of individuals were directly or indirectly involved in the preparation of this book. First, a number of my current and former students (S. Chatlani, L. K. Downie, J. C. Lapp, C. E. Lord, P. B. McGinnis, J. J. Noonan, and B. M. Wright) supplied unpublished data for many of the figures in this text. I am also grateful for the critical review of the text by S. Chatlani and B. Larrabee. I would like to offer special thanks to my mother, Jessie M. Shelby, and to my daughter, Stephanie R. Shelby, for their support of my efforts on this project throughout the past two years.

Chapter 1

Introduction

The presence of glasses in our everyday environment is so common that we rarely notice their existence. Our current casual attitude toward the family of materials known as glasses has not always existed. Early Egyptians considered glasses as precious materials, as evidenced by the glass beads found in the tombs and golden death masks of ancient Pharaohs. The cave dwellers of even earlier times relied on chipped pieces of obsidian, a natural volcanic glass, for tools and weapons, *i.e.*, scrapers, knives, axes, and heads for spears and arrows.

Humans have been producing glasses by melting of raw materials for thousands of years. Egyptian glasses date from at least 7000 BC. How did the first production of artificial glasses occur? One scenario suggests that the combination of sea salt (NaCl) and perhaps bones (CaO) present in the embers of a fire built on the sands (SiO_2) at the edge of a saltwater sea (the Mediterranean?) sufficiently reduced the melting point of the sand to a temperature where crude, low quality glass could form. At some later time, some other nomad found these lumps of glass in the sand and recognized their unusual nature. Eventually, some genius of ancient times realized that the glass found in the remains of such fires might be produced deliberately and discovered the combination of materials which led to the formation of the first commercial glasses.

The first crude manmade glasses were used to produce beads, or to shape into tools requiring sharp edges. Eventually, methods for production of controlled shapes were developed. Bottles were produced by winding glass ribbons around a mold of compacted sand. After cooling the glass, the sand was scraped from inside the bottle, leaving a hollow container with rough, translucent walls and usually lopsided shapes. Eventually, the concept of molding and pressing jars and bottles replaced the earlier methods and the quality of the glassware improved. It began to be possible to produce glasses which were reasonably transparent, although usually still filled with bubbles and other flaws.

The invention of glass blowing around the first century BC generated a greatly expanded range of application for glasses. The quality of glass jars and bottles improved dramatically, glass drinking vessels became popular, and the first clear sheet glasses became available, which eventually allowed the construction of buildings with enclosed windows. Colored glasses came into common use, with techniques for production of many colors regarded as family secrets, to be passed on from generation to generation of artisans. The method for producing red glasses by inclusion of gold in the melt, for example, was discovered and then lost, only to be rediscovered hundreds of years later. The combination of the discovery of many new colorants with the invention of glassblowing eventually lead to the magnificent stained glass windows of so many of the great cathedrals of Europe and the Near East.

The advent of the age of technology created many new opportunities for the application of glasses. The evolution of chemistry from the secretive practices of alchemists searching for the philosopher's stone to a profession involving millions of workers worldwide was strongly influenced by the invention of chemically resistant borosilicate glasses. Modern electronics became a reality with the invention of glass vacuum tubes, which evolved into the monitors for our computers and the televisions we watch every day. Recently, the development of glass optical fibers has revolutionized the telecommunication industry, with fibers replacing copper wires and radically expanding our ability to transmit flaw-free data throughout the world.

Unlike many other materials, glasses are also esthetically pleasing to an extent which far transcends their mundane applications as drinking vessels and ashtrays, windows and beer bottles, and many other everyday uses. Why are we so delighted with a lead crystal chandelier or a fine crystal goblet? Why do we find glass sculptures in so many art museums? Why are the stained glass windows of the great cathedrals so entrancing? What aspects of objects made of glass make them so desirable for their beauty as well as their more pragmatic uses?

The answer to these questions may lie in the ability of glasses to transmit light. Very few materials exist in Nature which are transparent to visible light. Metals are opaque, as are virtually all natural organic materials. Many liquids are transparent, but they are transient in nature, without the enduring qualities we desire in our possessions. A list of the few transparent natural solids includes diamonds, emeralds, rubies, and many other precious and semi-precious stones. It is difficult to think of a naturally transparent solid which is not highly valued for its transparency and brilliance. Our heritage as humans would seem to provide a bias toward placing a high value on such objects. We still are fascinated by 'bright, shiny objects'.

DEFINITION OF A GLASS

What is a 'glass'? The glasses used by mankind throughout most of our history have been based on silica. Is silica a required component of a glass? Since we can form an almost limitless number of inorganic glasses which do not contain silica, the answer is obviously 'no, silica is not a required component of a glass'. Glasses are traditionally formed by cooling from a melt. Is melting a requirement for glass formation? No, we can form glasses by vapor deposition, by *sol–gel* processing of solutions, and by neutron irradiation of crystalline materials. Most traditional glasses are inorganic and non-metallic. We currently use a vast number of organic glasses. Metallic glasses are becoming more common with every passing year. Obviously the chemical nature of the material cannot be used to define a glass.

What, then, is required in the definition of a glass? Every glass found to date shares two common characteristics. First, no glass has a long range, periodic atomic arrangement. Even more importantly, every glass exhibits the time-dependent behavior known as *glass transformation* behavior. This behavior occurs over a temperature range known as the glass transformation region. A glass can thus be defined as 'an amorphous solid completely lacking in long range, periodic atomic structure and exhibiting a region of glass transformation behavior'. Any material, inorganic, organic, or metallic, formed by any technique, which exhibits glass transformation behavior is a glass.

THE ENTHALPY/TEMPERATURE DIAGRAM

We have established that any material which exhibits glass transformation behavior is a glass. What, then, is glass transformation behavior? We traditionally discuss glass transformation behavior on the basis of either enthalpy or volume *versus* temperature diagrams, such as that shown in Figure 1.1. (This diagram will be discussed in considerably more detail in Chapter 6.) Since enthalpy and volume behave in a similar fashion, the choice of the ordinate is somewhat arbitrary. In either case, we can envision a small volume of a liquid at a temperature well above the melting temperature of that substance. As we cool the liquid, the atomic structure of the melt will gradually change and will be a characteristic of the exact temperature at which the melt is held.

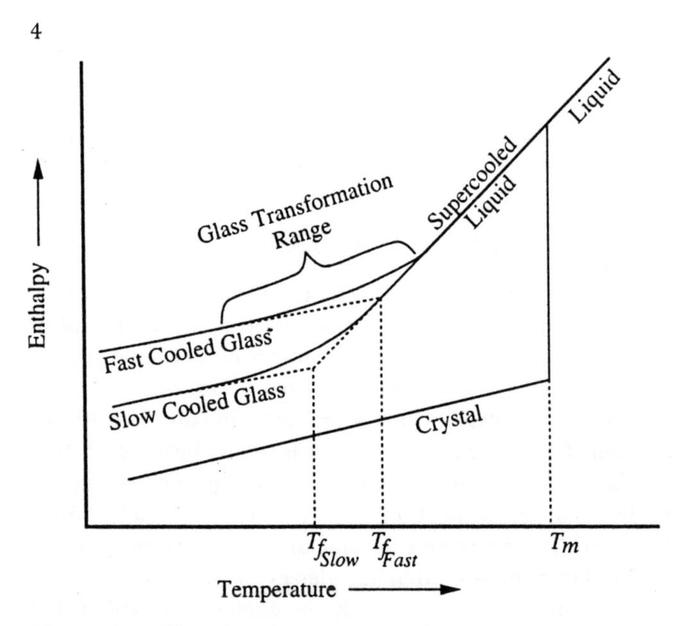

Figure 1.1 *Effect of temperature on the enthalpy of a glassforming melt*

Cooling to any temperature below the melting temperature of the crystal would normally result in the conversion of the material to the crystalline state, with the formation of a long range, periodic atomic arrangement. If this occurs, the enthalpy will decrease abruptly to the value appropriate for the crystal. Continued cooling of the crystal will result in a further decrease in enthalpy due to the heat capacity of the crystal.

If the liquid can be cooled below the melting temperature of the crystal without crystallization, a supercooled liquid is obtained. The structure of the liquid continues to rearrange as the temperature decreases, but no abrupt decrease in enthalpy due to a discontinuous structural rearrangement occurs. As the liquid is cooled further, the viscosity increases. This increase in viscosity eventually becomes so great that the atoms can no longer completely rearrange to the equilibrium liquid structure during the time allowed by the experiment. The structure begins to lag that which would be present if sufficient time were allowed to reach equilibrium. The enthalpy begins to deviate from the equilibrium line, following a curve of gradually decreasing slope, until it eventually becomes determined by the heat capacity of the frozen liquid, *i.e.*, the viscosity becomes so great that the structure of the liquid becomes fixed and is no longer temperature dependent. The temperature region lying between the limits where the enthalpy is that of the equilibrium liquid and that of the frozen solid is known as the *glass transformation region*. The frozen liquid is now a glass.

Since the temperature where the enthalpy departs from the equilibrium curve is controlled by the viscosity of the liquid, *i.e.*, by kinetic factors, use of a slower cooling rate will allow the enthalpy to follow the equilibrium curve to a lower temperature. The glass transformation region will shift to lower temperatures and the formation of a completely frozen liquid, or glass, will not occur until a lower temperature. The glass obtained will have a lower enthalpy than that obtained using a faster cooling rate. The atomic arrangement will be that characteristic of the equilibrium liquid at a lower temperature than that of the more rapidly cooled glass.

Although the glass transformation actually occurs over a temperature range, it is convenient to define a term which allows us to express the difference in thermal history between these two glasses. If we extrapolate the glass and supercooled liquid lines, they intersect at a temperature defined as the *fictive temperature.* The structure of the glass is considered to be that of the equilibrium liquid at the fictive temperature. Although the fictive temperature concept is not a completely satisfactory method for characterizing the thermal history of glasses, it does provide a useful parameter for discussion of the effect of changes in cooling rate on glass structure and properties. The changes which occur in the fictive temperature and the properties of glasses with subsequent reheating into the glass transformation region will be discussed in detail in Chapter 6.

Finally, we need to define a term, which, while commonly used, has only a vague scientific meaning. As indicated above, the glass transformation occurs over a range of temperatures and cannot be characterized by any single temperature. It is, however, convenient to be able to use just such a single temperature as an indication of the onset of the glass transformation region during heating of a glass. This temperature, which is termed either the *glass transformation temperature* (T_g) or the *glass transition temperature,* is rather vaguely defined by changes in either thermal analysis curves or thermal expansion curves. The values obtained from these two methods, while similar, are not identical. The value obtained for T_g is also a function of the heating rate used to produce these curves. Since T_g is a function of both the experimental method used for the measurement and the heating rate used in that measurement, it cannot be considered to be a true property of the glass. We can, however, think of T_g as a useful indicator of the approximate temperature where the supercooled liquid converts to a solid on cooling, or, conversely, where the solid begins to behave as a viscoelastic solid on heating. The utility of the concept of a glass transformation temperature will become much clearer in the following chapters.

Chapter 2

Principles of Glass Formation

INTRODUCTION

The earliest glasses used by man were found in nature. The ease of formation of sharp edges on obsidians, for example, allowed the production of knives, arrow heads, and other cutting tools. These naturally occurring glasses, which result from the cooling of molten rock, or lava, contain a wide variety of components, including alkali, alkaline earth, and transition metal oxides. In every case, however, silica is found to be the major constituent of these materials.

Since naturally occurring glasses proved to be so useful to early man, it is not surprising that the desire to produce glasses at will developed thousands of years ago. Furthermore, since all known glasses were silicates, it is also not surprising to find that the earliest man-made glasses were also silicates. In fact, very few non-silicate glasses were known prior to 1900. As a result, the first theories advanced to explain why some materials form glasses while others do not were based heavily on the existing knowledge of the behavior of silicate melts and the structure of silicate crystals. These theories tend to assume that some unique feature of certain melts leads to glass formation for those materials, while the lack of these features prevents the formation of glasses from other materials. These theories are often grouped under the heading of *structural theories of glass formation.*

In recent years, we have recognized the existence of a vast number of non-silicate glasses. In fact, we now know that polymers and metals can also readily be formed as glasses, as can a large number of non-oxide, inorganic compositions. We now recognize that virtually any material can be formed as a glass. Theories of glass formation no longer address the question of why a specific material will form a glass, but rather 'what do I have to do to make this material form a glass?' Since the emphasis has shifted from control of glass formation by selection of materials to control of glass formation by changes in processing, the importance of kinetics has become obvious. As a result, a new approach to glass formation, known as the *Kinetic Theory* of glass formation, has largely replaced the earlier structural theories.

STRUCTURAL THEORIES OF GLASS FORMATION

Perhaps the earliest, and simplest, theory of glass formation was based on the observation by Goldschmidt that glasses of the general formula R_nO_m form most easily when the ionic radius ratio of the cation, R, to the oxygen ion lies in the range between 0.2 and 0.4. Since radius ratios in this range tend to produce cations surrounded by four oxygen ions in the form of tetrahedra, Goldschmidt believed that only melts containing tetrahedrally coordinated cations form glasses during cooling. This contention was purely empirical, with no attempt to explain why tetrahedral coordination should be so favorable to glass formation.

A few years later, Zachariasen published a paper which extended the ideas of Goldschmidt and attempted to explain why certain coordination numbers might favor glass formation. Although intended only as an explanation for glass formation, this paper has become the basis for the most widely used models for glass structures and is probably the most widely cited (and frequently misquoted) paper in the inorganic glass literature. Essentially, Zachariasen noted that the silicate crystals which readily form glasses instead of recrystallizing after melting and cooling have network, as opposed to close-packed, structures. These networks consist of tetrahedra connected at all four corners, just as in the corresponding crystals, but the networks are not periodic and symmetrical as in crystals. These networks extend in all three dimensions, such that the average behavior in all directions is the same, *i.e.*, the properties of glasses are isotropic. Zachariasen contends that the ability to form such networks thus provides the ultimate condition for glass formation.

After establishing that the formation of a vitreous network is necessary for glass formation, Zachariasen considered the structural arrangements which could produce such a network. First, he contends that no oxygen atom can be linked to more than two network cations. Higher coordination numbers for the oxygen cations prevent the variations in oxygen–cation–oxygen bond angles necessary to form a non-periodic network. Zachariasen further noted that the only glasses known at the time of his work contained network cations in either triangular (B_2O_3) or tetrahedral (silicates, GeO_2, P_2O_5) coordination. He then generalized this observation by stating that the number of oxygen atoms surrounding the network cation must be small, specifically either three or four. Zachariasen clearly states that this requirement is empirical and is based solely on the lack of knowledge of any glasses in which the coordination number of the network cation is neither three nor four. The formation of a network in which the cations are located as far apart as possible further requires that the oxygen polyhedra be connected only at the corners and do not share either edges or faces. Finally, he states that the network can only be three-dimensional if at least three corners of each oxygen polyhedron are shared.

The thoughts of Zachariasen can be best summarized by his statement that formation of oxide glass *may* occur if (1) the material contains a high proportion of cations which are surrounded by either oxygen triangles or oxygen tetrahedra, (2) these polyhedra are connected only by their corners, and (3) some oxygen atoms are linked to only two such cations and do not form additional bonds with other cations. Essentially, item (1) states that sufficient network cations must be present to allow a continuous structure to form, item (2) states that the network is an open structure, and item (3) states that sufficient bonds linking the network polyhedra exist for the formation of a continuous network structure. It should be noted that Zachariasen only states that a glass *may* be formed under such conditions; he further states that the melt must be cooled under the proper conditions for glass formation actually to occur, thus anticipating the later theories based on the kinetics of the glass formation process.

A number of other statements by Zachariasen have become the basis for the models for glass structures termed the *random network theory*. These ideas will be discussed later under the topic of glass structure (Chapter 5). It is interesting to note, however, that the term 'random network' does not occur in the original work of Zachariasen, who referred to the glass structure as a 'vitreous network'. Furthermore, Zachariasen specifically states that the vitreous network is *not* entirely random due to the restriction of a minimum value for the internuclear distances. As a result, all internuclear distances are not equally probable and *X*-ray patterns of the type observed for glasses are a natural consequence of the vitreous network.

A number of other theories of glass formation are based on the nature of the bonds in the material. Smekal, for example, proposed that glasses are only formed from melts which contain bonds that are intermediate in character between purely covalent and purely ionic bonds. Since purely ionic bonds lack any directional characteristics, highly ionic materials do not form network structures. On the other hand, highly covalent bonds tend to force sharply defined bond angles, preventing the formation of a non-periodic network. Glassforming substances thus fall into the categories of either inorganic compounds which contain bonds which are partially ionic and partially covalent, or either inorganic or organic compounds which form chain structures, with covalent bonds within the chains and van der Waals bonds between the chains.

Stanworth attempted to quantify the mixed bond concept by use of the partial ionic character model of Pauling. He classified oxides into three groups on the basis of the electronegativity of the cation. Since the anion is oxygen in every case, this approach is effectively identical to grouping by fractional ionic character of the cation–anion bond. Cations which form bonds with oxygen with a fractional ionic character near 50% should act as *network formers* (group I) and produce good glasses. Cations with slightly lower electronegativities (group II), which form slightly more ionic bonds with oxygen, cannot form glasses by themselves, but can partially replace cations from the first group. Since these ions behave in a manner which is intermediate between that of cations which do form glasses and those which never form glasses, they are known as *intermediates*. Finally, cations which have very low electronegativities (group III), and therefore form highly ionic bonds with oxygen, never act as network formers. Since these ions only serve to modify the network structure created by the network-forming oxides, they are termed *modifiers*.

Bond strength has also been used as a criterion for predicting ease of glass formation. Sun argued that strong bonds prevent reorganization of the melt structure into the crystalline structure during cooling and thus promote glass formation. In this particular case, bond strength was defined as the energy required to dissociate an oxide into its component atoms in the gaseous state. Since experimental values for this energy include contributions from all cation–anion bonds, the strength of a single bond is determined by dividing the dissociation energy by the number of cation–anion bonds in the coordination unit, *e.g.*, four for a tetrahedrally coordinated silicon ion. Use of this criterion yields results similar to those of Stanworth, with groups of network former, intermediate, and modifier cations. Although this model yields results which are compatible with empirical observations, it has not proven to yield any particular insight into the process of glass formation.

Finally, Rawson (1967) suggests that Sun ignored the importance of temperature in his model. He suggests that high melting temperatures mean that considerable energy is available for bond disruption, while low melting temperatures mean that significantly less energy is available. It follows that a material with a large single bond strength and a low melting temperature will be a much better glassformer than one with a similar single bond strength but a much higher melting temperature. Application of this model to single cation oxides does little to increase the information obtained using the Sun model, although it does predict the extremely good glassforming character of boric oxide. Extension to binary and ternary systems, however, yields the prediction that the ease of glass formation should be improved for compositions near eutectics in binary and ternary systems. This phenomena has frequently been observed and is called the 'liquidus temperature effect'. Glass formation in the $CaO–Al_2O_3$ binary in a region near a eutectic is often cited as an example of this effect.

KINETIC THEORIES OF GLASS FORMATION

The theories discussed above consider only the relative ease of glass formation. Any compound or mixture which forms a glass during cooling from the melt at a moderate cooling rate is considered to be a 'good' glassformer, while materials which require a more rapid cooling rate in order to form a glass are considered to be 'poor' glassformers. Melts which cannot be cooled to form glasses without uses of extreme cooling rates are considered to be non-glassformers.

More recent theories regarding glass formation have reconsidered the criteria for defining a substance as a glassformer. It is now recognized that virtually any material will form a glass if cooled so rapidly that insufficient time is provided to allow the reorganization of the structure into the periodic arrangement required by crystallization. It follows that the question is not *whether* a material will form a glass, but rather *how fast* it must be cooled to avoid detectable crystallization.

Nucleation

The term crystallization actually refers to a combination of two processes: *nucleation* and *crystal growth.* Crystallization requires the presence of a nucleus (nucleation) on which the crystal will subsequently grow (crystal growth) to a detectable size. The nucleus may be either *homogeneous*, *i.e.*, forming spontaneously within the melt, or *heterogeneous*, *i.e.*, forming at a pre-existing surface such as that due to an impurity, crucible wall, etc. If no nuclei are present, crystal growth cannot occur and the material will form a glass. Even if some nuclei are present, but no growth has occurred, the extremely small size and low volume fraction of the nuclei prevents their detection, so that the solid is, for all practical purposes, still a glass.

Classical nucleation theory addresses the process of homogeneous nucleation, where nuclei are formed with equal probability throughout the bulk of the melt. Nuclei are extremely small and, at least for the case of homogeneous nucleation, are not usually detected directly. Their concentration is usually determined by a complex experiment involving an isothermal heat treatment at the nucleation temperature, quenching the sample to freeze in the nuclei, and then reheating the sample to a temperature where the nuclei can grow to a detectable size for analysis. Since the assumption that no change in nuclei concentration occurs during the quenching or crystal growth stages is always questionable, these experiments can result in considerable error. The nucleation rate, I (number of nuclei per unit volume formed per unit time), is then determined by dividing the concentration of nuclei by the total time of the isothermal heat treatment at the nucleation temperature.

Derivations of the equation expressing the nucleation rate as a function of a number of factors can be found in many sources. Comparison of these derivations can be somewhat confusing due to differences in terminology, use of different symbols for the same quantity, and differing units. Essentially, however, all of these derivations begin with the formation of a spherical nucleus. Two barriers exist to the formation of a nucleus. First, the thermodynamic barrier involves the free energy change in a system when a nucleus is formed. The second, or kinetic, barrier is the result of the requirement that mass be moved or rearranged in space to allow the growth of an ordered particle (a crystal) from a disordered liquid. The overall process is described by the expression

$$I = A\exp[-(W^* + \Delta G_D)/kT] \tag{2.1}$$

where A is a constant, W^* and ΔG_D are the thermodynamic and kinetic free energy barriers to nucleation, respectively, k is the Boltzmann constant, and T is the absolute temperature (K). In this case, W^* is actually the work required to form a nucleus of critical size, *i.e.*, one which will grow instead of redissolve into the melt. The pre-exponential constant A in Equation 2.1 is given by

$$A = 2n_v V^{1/3}(kT/h)(\gamma/kT)^{1/2} \tag{2.2}$$

where n_v is the number of formula units of the crystallizing component phase per unit volume of the melt, V is the volume per formula unit, γ is the crystal–melt interfacial free energy per unit area, and h is Planck's constant. James notes that A is essentially constant over the temperature range of nucleation measurement and, to a good approximation, can be expressed by

$$A = n_v(kT/h) \tag{2.3}$$

Formation of a nucleus actually involves two changes in the energy of the system, *i.e.*, the thermodynamic barrier. First, the formation of a crystalline arrangement will lower the volume free energy since the crystalline state has a lower free energy than the melt. This decrease in free energy is countered by an increase in surface energy due to the formation of a new interface between regions of different structures. The net change in energy, W, for a sphere of radius r is given by the sum of these terms, or

$$W = \frac{4}{3}\pi r^3 \Delta G_v + 4\pi r^2 \gamma \tag{2.4}$$

where the first term represents the change in volume free energy (ΔG_v is the change in volume free energy per unit volume) and the second term represents the change in surface energy. Since ΔG_v is negative for temperatures below the melting point (T_m) of the crystal, the terms have opposite signs. Since nuclei are small, the surface energy term will dominate at very low values of r, W will increase with increasing r, and the nucleus will be unstable. If, however, the nucleus can survive to grow to a large enough size, the first term will become larger than the second, W will begin to decrease with increasing nucleus size, and the nucleus will become stable.

If we take the derivative of W with respect to r and set it equal to zero, we can determine the value for r where the nucleus just becomes stable, *i.e.*

$$\frac{dW}{dr} = 4\pi r^2 \Delta G_v + 8\pi r \gamma = 0 \tag{2.5}$$

This value of r, which is known as the critical radius, r^*, is then given by

$$r^* = -2\gamma/\Delta G_v \tag{2.6}$$

If we now substitute r^* into Equation 2.4, we can define the value of W for the critical nucleus, W^*, by the expression

$$W^* = 16\pi\gamma^3/3\Delta G_v^2 \tag{2.7}$$

If we express the free energy change per mole, ΔG, instead of per unit volume, as is sometimes done, we can make the substitution

$$\Delta G = V_m \Delta G_v \tag{2.8}$$

and rewrite Equation 2.4 in the form

$$W^* = 16\pi\gamma^3 V_m^2/3\Delta G^2 \tag{2.9}$$

where V_m is the molar volume of the crystal phase and ΔG is the bulk free energy change per mole in crystallization.

A number of approximations can be made for the value of ΔG or ΔG_v. Since $\Delta G = V_m \Delta G_v$, expressions can be written on either a molar or volume basis, *i.e.* either as

$$\Delta G = -\Delta H_f(T_m - T)/T_m \tag{2.10}$$

where ΔH_f is the heat of fusion, or as

$$\Delta G_v = -\Delta H_f(T_m - T)/V_m T_m \tag{2.11}$$

ΔG is also sometimes written as

$$\Delta G = -\Delta H_f(T_m - T)T/T_m^2 \tag{2.12}$$

Since these expressions are only applicable for small undercoolings, *i.e.*, when $T_m - T$ is small, the exact form of the expression used is not critical. The errors introduced by the use of these approximations increase as the degree of undercooling increases.

Determination of the kinetic barrier for nucleation, ΔG_D, is subject to some controversy. Obviously, this barrier can be discussed in terms of an effective diffusion coefficient, D, which is given by

$$D = (kT\lambda^2/h)\exp(-\Delta G_D/kT) \tag{2.13}$$

where λ is the atomic jump distance. In many cases, it has been assumed that D is related to the viscosity, η, of the melt *via* the Stokes–Einstein relation

$$D = kT/3\pi\lambda\eta \tag{2.14}$$

Using the relations given by Equations 2.13 and 2.14, and substituting into our original expression, Equation 2.1, we obtain the expression

$$I = (Ah/3\pi\lambda^3\eta)\exp(-W^*/kT) \tag{2.15}$$

If we further substitute the simplified expression for A, as given by Equation 2.3, into Equation 2.15, we can also write this expression as

$$I = (n_v kT/3\pi\lambda^3\eta)\exp(-W^*/kT) \tag{2.16}$$

These expressions are derived for the case of homogeneous nucleation.

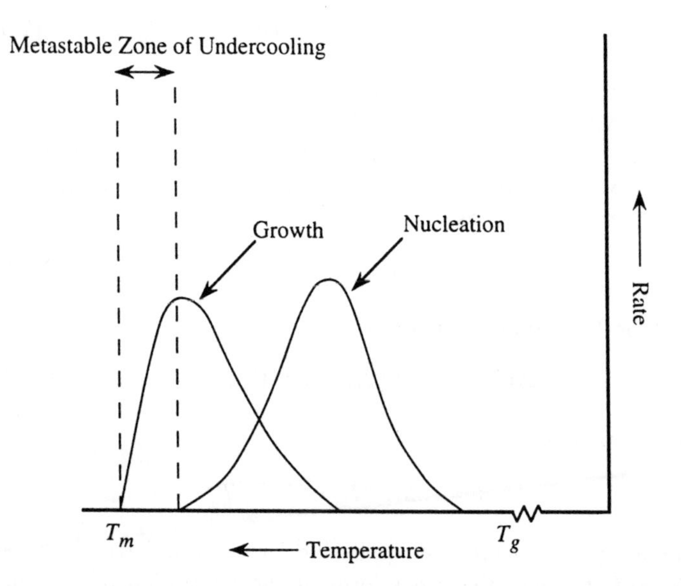

Figure 2.1 *Effect of temperature on the rates of nucleation and crystal growth for a glassforming melt*

James (1982) indicates that an expression similar to Equation 2.1 can be used for heterogeneous nucleation on a flat substrate. In this case

$$I_{het} = A_{het}\exp[-(W^*_{het} + \Delta G_D)/kT] \tag{2.17}$$

where the subscript 'het' refers to values for heterogeneous nucleation and W^*_{het} is a function of the angle of contact between the crystal nucleus and the substrate. If we express A_{het} by an expression similar to Equation 2.3

$$A_{het} = n_s(kT/h) \tag{2.18}$$

where n_s is the number of formula units of the melt in contact with the substrate per unit area, we arrive at the expression for the heterogeneous nucleation rate, given by

$$I_{het} = n_s(kT/h)\exp[-(W^*_{het} + \Delta G_D)/kT] \tag{2.19}$$

The expressions derived above can be used to predict the shape of the curve for the nucleation rate, I, *versus* temperature. Since glasses are usually formed during cooling from a melt initially held at a temperature above T_m, discussion of the effect of temperature on the nucleation rate traditionally follows the same path, *i.e.*, from higher toward lower temperature. This process is illustrated schematically in Figure 2.1.

As long as the melt is held at a temperature above T_m, there will be no tendency toward formation of a nucleus. As the temperature is lowered to $\leq T_m$, the change in the free energy of the system is such that a nucleus can become stable. If the temperature is very near T_m, however, the value of ΔG_v is very small. It follows that the critical radius for a stable nucleus, as given by Equation 2.6, will be very large. Since the probability of a nucleus reaching such a large size is extremely low, the melt will remain effectively free of nuclei, even though the temperature is below T_m. As the temperature decreases further, ΔG_v will increase, thus decreasing the value of the critical radius. Eventually the critical radius will become so small (often only a few tenths of a nanometer) that the probability of formation of a nucleus large enough to exceed the critical radius will become significant and nuclei will begin to be formed in detectable quantities. Since a significant degree of undercooling, which may be as little as a small fraction of a degree or as much as a few hundred degrees, must occur before the critical radius decreases to the point where detectable nuclei exist in the melt, a metastable zone of undercooling exists.

Once the temperature of the cooling melt passes below the lower limit of the metastable zone, the thermodynamic barrier will decrease with decreasing temperature, allowing nuclei to form at an ever increasing rate. If the viscosity of the melt is low, there will be little kinetic obstruction to nucleus formation and the nucleation rate will increase rapidly with decreasing temperature as ΔG_v increases (see Equation. 2.10, 2.11, and 2.12). We must remember, however, that the viscosity is also highly temperature dependent, so that the kinetic barrier to nucleation will also increase rapidly with decreasing temperature. As the kinetic barrier increases, it will eventually force the nucleation rate to begin to decrease and eventually fall to essentially zero. The conflicting changes in the nucleation rate due to changes in the thermodynamic and kinetic barriers thus will result in a maximum in the temperature dependence of the nucleation rate.

Crystal Growth

A large number of expressions describing crystal growth can also be found in the literature. Many of these equations deal with specific models for different crystal growth mechanisms. A simple general model, however, can be derived using arguments similar to those used for the nucleation rate. In this case, a general equation for the crystal growth rate, U, is given by the expression

$$U = a_0 \nu \exp(-\Delta E/kT)[1 - \exp(\Delta G/kT)] \tag{2.20}$$

where a_0 is the interatomic separation distance, ν is the vibrational frequency, and ΔE and ΔG are the kinetic and thermodynamic barriers to crystal growth. If we use the same arguments as for nucleation rate, and introduce Equations 2.13 and 2.14 to describe diffusion, we obtain the expression

$$U = \left(\frac{kT}{3\pi a_0^2 \eta}\right)[1 - \exp(\Delta G/kT)] \tag{2.21}$$

Further substitutions of equations of the form of Equation 2.10 or 2.12 allow the replacement of ΔG by ΔH_f in Equation 2.21.

The temperature dependence of the crystal growth rate, as expressed by Equation 2.21, is very similar to that for the nucleation rate, as is shown in Figure 2.1. The principal difference lies in the lack of a metastable zone for crystal growth. Since growth can occur at any temperature below T_m so long as a nucleus is available, and that nucleus need not have formed during cooling, detectable growth rates can occur at any temperature $<T_m$. The nuclei involved need not even have the same composition as that of the growing crystal, which is frequently the case for heterogeneous nucleation, particularly at surfaces. Once again, if the viscosity is low, the growth rate will be determined by the thermodynamic values and will tend to be large. As the temperature decreases, the viscosity will increase rapidly, slowing and eventually halting crystal growth. The resulting curve of U *versus* temperature will exhibit a maximum and eventually decrease to zero at lower temperatures.

General Kinetic Treatment of Glass Formation

Formation of a glass involves cooling from the melt in such a manner as to prevent significant crystal formation. The models for nucleation and crystal growth rates discussed above treat nucleation and crystal growth as independent entities. In reality, however, nucleation and crystal growth occur simultaneously during cooling of a melt, with rates which change continuously as the temperature decreases. A pragmatic approach to glass formation must, therefore, deal with the interaction between these processes. Under these conditions, we can describe the volume fraction of crystals in a sample, V_x/V, where V_x is the volume of crystals and V is the sample volume, by the expression

$$\frac{V_x}{V} = 1 - \exp\left[-\int_0^t I_v\left(\int_{t'}^t U\,d\tau\right)^3 dt'\right] \tag{2.22}$$

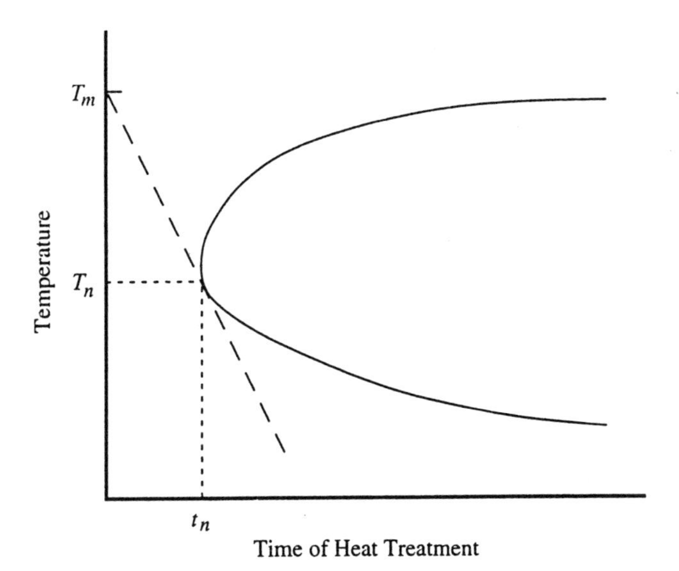

Figure 2.2 *A time-temperature-transformation curve for a glassforming melt*

where I_v is the nucleation rate per unit volume and U is the linear crystal growth rate. During cooling, both I_v and U will be time dependent due to their temperature dependence.

Under isothermal conditions, Equation 2.22 can be simplified to the form

$$\frac{V_x}{V} = 1 - \exp\left(-\frac{\pi}{3} I_v U^3 t^4\right) \tag{2.23}$$

where t is the time the sample has been held at the experimental temperature. If I_v and U are known at a given temperature, one can calculate the time required to form a specific volume fraction of crystals. Furthermore, if these quantities are known as a function of temperature, one can calculate the curve in temperature/time space which corresponds to that specific value of V_x/V. This curve is known as a time–temperature–transformation, or TTT, curve. As a result of the competition between thermodynamic and kinetic factors for both the nucleation and growth rates, this curve will have the general shape shown in Figure 2.2. Since I_v and U approach zero as the temperature approaches T_m, the time required to form the specified volume fraction of crystals will approach infinity. At very low temperatures, the values of I_v and U also approach zero due to the very high viscosity of the melt, and the time to reach the specified value of V_x/V also approaches infinity. The least favorable conditions for glass formation occur at the temperature, corresponding to the 'nose' of the curve, T_n where the time to reach the curve, t_n, is the least.

All combinations of heat treatment times and temperatures to the left of this curve will yield samples containing less than our specified volume fraction of crystals, while any combination of time and temperature to the right of this curve will yield a larger volume fraction of crystals. If we now define samples containing less than some arbitrary volume fraction of crystals as glasses (often 1 ppm, or 10^{-6}), we now know what experimental conditions we must satisfy to form a glass for that particular material. The critical cooling rate, *i.e.*, the minimum cooling rate required to yield a glass, can be obtained from the slope of the tangent to the curve, when the initial conditions are defined as T_m at time zero. (Since crystallization cannot occur at temperatures above T_m, the time when the cooling melt reaches T_m is obviously the time when the experiment effectively begins.) This critical cooling rate, $(dT/dt)_c$, is given by the expression

$$\left(\frac{dT}{dt}\right)_c \approx \frac{(T_m - T_n)}{t_n} \tag{2.24}$$

If we know the thermal diffusivity, D_T, of the melt, the maximum, or critical, thickness, L_c, obtainable as a glass can be estimated from the expression

$$L_c \approx \sqrt{(D_T t_n)} \tag{2.25}$$

If we change our criteria for defining a sample as a glass, the critical cooling rate required to meet those criteria will obviously change. Since TTT curves for greater values of V_x/V will lie to the right of the curve shown in Figure 2.2, the critical cooling rate will decrease as we increase the allowable volume fraction of crystals in the sample. This volume fraction is usually quite arbitrary, since few samples are examined at the level necessary to detect the difference between a sample containing a volume fraction of crystals of 1 ppm and one containing, for example, a volume fraction of 100 ppm. The crystal content which might be acceptable for a window glass will be quite different from that which is acceptable for an optical fiber, a lens, or the face plate of a video display. The location of the crystals within the sample, as well as their size distribution, clearly affect our ability to detect them. A single, large crystal at the surface of a sample presents a completely different situation from that for an equal volume of very small crystals distributed throughout the bulk of the sample. As a result, the TTT diagram must be considered as a model for aiding our understanding of the glass formation process rather than as an experimental tool. At this time, these diagrams have been produced for very few materials.

Although TTT diagrams are not widely used, the general kinetic approach does provide considerable insight into the criteria for the formation of glasses. First, this approach alters our thinking from questions regarding why certain melts readily form glasses while others do not, to questions regarding the conditions necessary to force a melt to form a glass during cooling. The concept of a critical cooling rate thus leads to the simple question of HOW FAST must a sample be cooled to form a glass. We can now use this approach to consider what factors will lead to small critical cooling rates and how these factors relate to the concepts originally derived on the basis of melt structure and bonding.

Melt viscosity is clearly an important factor in glass formation. The kinetic barrier to crystallization will be very large if the melt viscosity is large at T_n. A high viscosity at T_n can result from either a high viscosity at the melting temperature, as is found for silica, boric oxide, and many other oxide glasses, or from a very steep viscosity/temperature curve, such that the viscosity, while low at T_m, rises to a high value by T_n. Typical organic glassformers and many of the newer glasses formed in non-oxide systems, *e.g.*, fluoride and other halide glasses, fall into the latter category. In general, complex melt structures containing large units, which are typical of melts of crystals with network structures, lead to high viscosities at the melt temperature. Since network structures are common in silicate, borates, germanates, and other oxides with small coordination numbers for the principal cation, the structural model of Zachariasen correctly predicts the good glassforming behavior of such materials. In many cases, the combination of a high viscosity at T_m and a steep viscosity/temperature curve is favored by a low T_m, which explains the observation that glass formation is favored in regions of eutectics in binary and ternary systems.

Melts which exhibit a large barrier to nucleation also exhibit good glassforming behavior. These materials typically have a large crystal–melt surface energy and a large entropy of fusion. Inhibition of crystal growth due to factors other than simply high viscosity also improves glassforming behavior. Complex melt structures obviously inhibit rearrangement of the melt into the ordered crystalline structure. Such complexity occurs naturally for many melts, or can be enhanced by the 'kitchen sink' method, *i.e.*, use many different elements in the composition, so that the redistribution of ions to the appropriate sites on the growing crystals is more difficult. This approach is routinely used in commercial glass technology and partially explains the complex compositions of many common glasses.

Glass formation can also be influenced by external factors. Since growth of crystals on heterogeneous nuclei does not require formation of homogeneous nuclei, a melt which is free of potential heterogeneous nuclei can be cooled more easily to form a glass than can a melt which contains a large concentration of such nuclei. Elimination of heterogeneous nuclei prevents crystal growth in the region of metastable nucleation, since no nuclei will then exist in this region. Elimination of these nuclei is enhanced by the addition of strong fluxes to a batch, which partially explains the strong effect of PbO in enhancing the ease of glass formation, by superheating the melt, by elimination of impurities such as refractory particles, and by minimizing contact with container walls. Although the first three conditions are routinely met by changes in composition or processing, production of glasses using containerless melting is much more difficult. Approaches to containerless melting range from microgravity studies in the space shuttle and drop towers, to various schemes for levitation of melts, to use of rapid cooling methods such as plasma spraying, where the melt is never actually in contact with a container.

DETERMINATION OF GLASSFORMING ABILITY AND GLASS STABILITY

Formation of a glass is a rather simple process. The appropriate batch is prepared, placed in a crucible, heated to form a crystal-free melt, and cooled to room temperature. The sample is examined to determine if it contains crystals, using methods ranging from casual visual examination to *X*-ray or electron diffraction. If no crystals are detected, the sample is deemed to be a glass; if crystals are detected, it is described as either partially or fully crystallized, depending upon the extent of crystallization.

Quantitative comparison of the glassforming ability of closely related compositions is not nearly so easily performed. Both compositions may form a glass under the specific conditions used in a simple experiment, but may display significantly different tendencies toward glass formation under other conditions. Small changes in melt size, for example, may result in crystallization of one melt, while the other still forms a glass. Some calcium aluminate glasses, for example, may be readily formed from 5 g melts, while they cannot be formed without rapid quenching from 10 g melts. A very slight change in the CaO to Al_2O_3 ratio will allow the routine formation of glasses from 10 g melts. How do we now characterize the difference in the glassforming ability of these closely related compositions?

The answer to our problem is provided by the general kinetic approach to glass formation. The ease of glass formation can be defined by the critical cooling rate required to prevent crystallization of a specified volume fraction of the sample. If we determine the critical cooling rate for two melts of identical size, the melt with the smaller critical cooling rate has the better glassforming ability. A series of compositions can thus be ranked in order of improving glassforming ability by decreasing values of their critical cooling rates.

Unfortunately, while the concept of a critical cooling rate is easy to understand, the actual measurement of these rates is rather tedious. A melt must be heated to a pre-determined temperature, held for a specified time to completely dissolve any residual crystals and nuclei, and then cooled at a linear rate until it forms a solid, which is then examined to determine if it is a glass. If so, the sample, or a new, identical batch, is heated to the melt temperature and the experiment is repeated using a slower cooling rate. This procedure is repeated until the sample fails to form a glass. The slowest cooling rate which produced a glass is then defined as the critical cooling rate. Although only two cooling experiments are required to bracket the critical cooling rate, actual practice will usually require several experiments for each composition.

Determination of critical cooling rates can sometimes be carried out using a differential scanning calorimeter (DSC). In this case, the process can be automated, with a computer repeatedly carrying out the experiment with a series of decreasing cooling rates. Cooling rates below the critical will often be characterized by an exothermic peak in the thermal spectrum due to the release of the heat of crystallization. Absence of such a peak in a thermal spectrum is often taken as evidence of a lack of crystal formation, *i.e.*, glass formation. This technique is best suited for low melting temperature batches which readily crystallize with a large exothermic effect if cooled at a rate less than the critical cooling rate.

While glassforming ability is defined in terms of resistance to crystallization of a melt during cooling, glass stability is defined in terms of resistance to crystallization of a glass during heating. Glassforming ability is most important during processes requiring production of an initial glass, while glass stability is most important during processes involving reforming of an existing glass. As an example, consider the preparation of optical fibers by drawing from a preform. A preform consists of a cylinder of a core glass surrounded by a cladding glass. The entire preform is heated to a drawing temperature. Fibers are drawn directly from the preform. Crystals formed during cooling to produce the preform provide pre-existing flaws which may be carried over to the optical fiber. Crystals formed during reheating to the fiber drawing temperature will provide similar flaws. Since the crystalline phases formed during cooling may differ from those formed during reheating, it may be possible to differentiate between flaws due to poor glassforming ability and to those due to poor glass stability.

Although these two properties are not identical, they are frequently confused in the literature and in technological practice. It is often assumed that poor glassforming ability automatically leads to poor glass stability and *vice versa*. This is not necessarily true, however. Calcium aluminate glasses, for example, exhibit poor glassforming ability, but, once formed, exhibit very reasonable glass stability. Many other compositions exhibit similar behavior.

Glass stability is frequently characterized by the difference in temperature between the onset of the glass transformation region (T_g) and the occurrence of crystallization (T_x) for a sample heated at a specified linear rate. These measurements are routinely carried out using a differential scanning calorimeter or differential thermal analyzer (DTA). The exact definitions of T_g and T_x are subject to the preference of the experimenter, as is the choice of the appropriate heating rate used in the study. In particular, some choose to define T_x as the temperature of the first maximum in the thermal spectrum, while others define T_x as the extrapolated onset temperature for the first crystallization exotherm. Others argue that the quantity $T_x - T_g$ should be normalized by T_g, T_x, or T_m of the crystalline phase to compare behavior of glasses which crystallize in very different temperature ranges. As a result, there is no universally accepted criteria for glass stability. In general, however, so long as samples are compared using identical criteria, most studies will yield similar results.

Typical thermal spectra (see Figures 2.3 and 2.4) may contain one or more exothermic peaks due to crystallization of different phases, but only the lowest temperature peak is considered in discussing glass stability. Once a significant number of crystals are formed, subsequent events at higher temperatures are not considered important in glass stability discussions.

Since glasses which are relatively stable, including almost all common commercial compositions, crystallize so slowly that they will not exhibit a crystallization exotherm during heating at the usual 10 or 20 K min^{-1} used in these studies, they cannot be characterized by this method. In these cases, it is usually necessary to carry out a series of isothermal heat treatments in order to determine the conditions under which stable glasses crystallize to a detectable extent. Creation of a TTT diagram

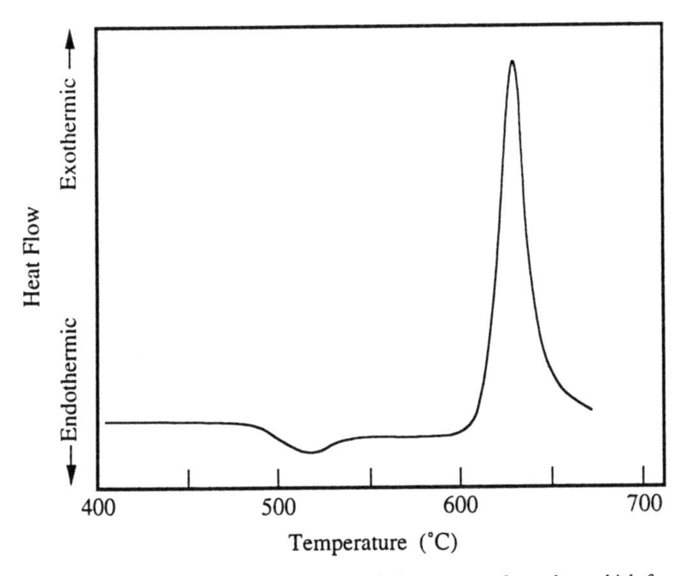

Figure 2.3 *A typical differential scanning calorimeter curve for a glass which forms a single crystalline phase on reheating.*

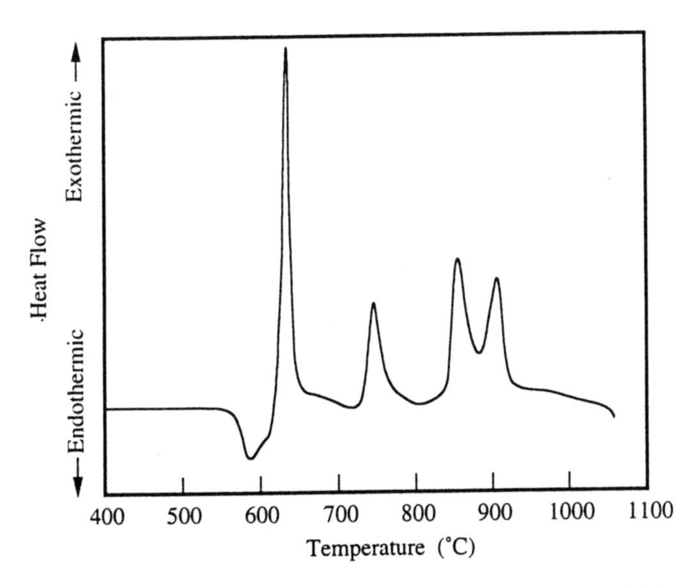

Figure 2.4 *A typical differential scanning calorimeter curve for a glass which forms several crystalline phases on reheating.*

often proves to be an ideal method for describing the stability of such glasses. In other cases, determination of crystal growth rates for related glasses as a function of temperature may be used to compare the stability of glasses which are highly resistant to crystallization.

SUMMARY

Virtually any material can be formed as a glass under the proper experimental conditions. Kinetic theories describing nucleation and crystal growth rates allow the prediction of these conditions. Examination of the results of kinetics studies allows more general statements regarding the effect of crystal structures and the structures of the melts derived from these crystals on the ease of formation of glasses from melts. These general trends correlate quite well with earlier theories of glass formation which were based on structural and bonding considerations. In particular, the temperature dependence of the viscosity is shown to be a major factor in determining the glassforming ability of a given melt. The use of TTT curves and the concept of critical cooling rate provide direct experimental methods for comparisons of glassforming ability among various materials.

Chapter 3

Glass Melting

INTRODUCTION

Although glasses can be made by a wide variety of methods, the vast majority are still produced by melting of batch components at an elevated temperature. This procedure always involves the selection of raw materials, calculation of the relative proportions of each to use in the batch, and weighing and mixing these materials to provide a homogeneous starting material. During the initial heating process, these raw materials undergo a series of chemical and physical changes to produce the melt. Conversion of this melt to a homogeneous liquid may require further processing, including the removal of any unmelted batch remnants, impurities, and bubbles. Production of commercial products requires forming of specific shapes as well as heat treatments to remove stresses generated during the cooling process or to produce glasses strengthened by thermal tempering.

RAW MATERIALS

In general, glasses are either produced from high-quality, chemically pure components or from a mixture of far less pure minerals. Research specimens, optical glasses, and many glasses used for low-volume, high-technology applications are produced using those chemicals we might routinely encounter in any chemical laboratory. Bulk commercial products, on the other hand, are produced from minerals, which typically have names and compositions which are not familiar to the novice. The names of many of these minerals and their compositions are listed in Table 3.1. Gravimetric factors, which allow calculation of the yield of the desired glass component for each weight unit of raw material, are also listed in this table.

Table 3.1 *Raw Materials for Glassmaking*

Common Name	*Nominal Composition*	*Gravimetric Factor**
Albite feldspar	$Na_2O–Al_2O_3–6SiO_2$	Na_2O = 8.46 Al_2O_3 = 5.14 SiO_2 = 1.45
Alumina	Al_2O_3	Al_2O_3 = 1.00
Alumina hydrate	$Al_2O_3·3H_2O$	Al_2O_3 = 1.53
Anorthite feldspar	$CaO-Al_2O_3–2SiO_2$	CaO = 4.96 Al_2O_3 = 2.73 SiO_2 = 2.32
Aplite	Alkali lime feldspar	Varies with exact composition
Aragonite	$CaCO_3$	CaO = 1.78
Bone ash	$3CaO–P_2O_5$ or $Ca_3(PO_4)_2$	CaO = 1.84 P_2O_5 = 2.19
Barite (barytes) (Heavy spar)	$BaSO_4$	BaO = 1.52
Borax	$Na_2O–2B_2O_3·10H_2O$	Na_2O = 6.14 B_2O_3 = 2.74
Anhydrous borax	$Na_2O–2B_2O_3$	Na_2O = 3.25 B_2O_3 = 1.45
Boric acid	$B_2O_3·3H_2O$	B_2O_3 = 1.78
Burnt dolomite	CaO–MgO	CaO = 1.72 MgO = 2.39
Caustic potash	KOH	K_2O = 1.19
Caustic soda	NaOH	Na_2O = 1.29
Cryolite	$3NaF–AlF_3$	NaF = 1.67 AlF_3 = 2.50
Cullet	Scrap glass	Varies with exact composition
Dolomite	$CaCO_3–MgCO_3$	CaO = 3.29 MgO = 4.58
Fluorspar	CaF_2	CaF_2 = 1.00
Gypsum	$CaSO_4·2H_2O$	CaO = 3.07
Kyanite	$Al_2O_3–SiO_2$	Varies with exact composition
Lime (quick lime) (Burnt lime)	CaO	CaO = 1.00
Limestone (calcite)	$CaCO_3$	CaO = 1.78
Litharge (yellow lead)	PbO	PbO = 1.00
Microcline	$K_2O–Al_2O_3–6SiO_2$	K_2O = 5.91 Al_2O_3 = 5.46 SiO_2 = 1.54
Nepheline	$Na_2O–Al_2O_3–2SiO_2$	Na_2O = 2.84 Al_2O_3 = 1.73 SiO_2 = 1.47

Nepheline syenite	Mixture of nepheline and feldspars	Varies with exact composition
Niter (saltpeter)	KNO_3	$K_2O = 2.15$
Potash	K_2O or	$K_2O = 1.00$
	K_2CO_3	$K_2O = 1.47$
Red lead	Pb_3O_4	$PbO = 1.02$
Salt cake	Na_2SO_4	$Na_2O = 2.29$
Sand (Glassmaker's sand) (Potter's flint)	SiO_2	$SiO_2 = 1.00$
Slag	Blast furnace waste glass	Varies with exact composition
Slaked lime	$CaO{\cdot}H_2O$ or $Ca(OH)_2$	$CaO = 1.32$
Soda ash	Na_2CO_3	$Na_2O = 1.71$
Soda niter (Chile saltpeter)	$NaNO_3$	$Na_2O = 2.74$
Spodumene	$Li_2O–Al_2O_3–4SiO_2$	$Li_2O = 12.46$ $Al_2O_3 = 3.65$ $SiO_2 = 1.55$
Whiting	$CaCO_3$	$CaO = 1.79$

* Quantity required to yield one weight unit of the glass component.

Regardless of the source of the components used to produce a specific glass, the batch materials can be divided into five categories on the basis of their role in the process: *glassformer*, *flux*, *property modifier*, *colorant*, and *fining agent*. The same compound may be classed into different categories when used for different purposes. Alumina, for example, serves as a glassformer in aluminate glasses, but is considered a property modifier in most silicate glasses. Arsenic oxide may be either a glassformer or a fining agent, depending upon the purpose for which it has been added to the batch.

The most essential component of any glass batch is always the glassformer. Every glass contains one or more components which serve as the primary source of the structure. While these components are commonly designated as glassformers, they are also called *network formers* or *glassforming oxides* in many oxide glasses. The identity of these components usually serves as the basis for the generic name used for the glass. If most of the glassformer present in a specific sample is silica, for example, that glass is called a silicate. If a significant amount of boric oxide is also present in addition to silica, the sample is termed a borosilicate glass.

The primary glassformers in commercial oxide glasses are silica (SiO_2), boric oxide (B_2O_3), and phosphoric oxide (P_2O_5), which all readily form single component glasses. A large number of other compounds may act as glassformers under certain circumstances, including GeO_2, Bi_2O_3, As_2O_3, Sb_2O_3, TeO_2, Al_2O_3, Ga_2O_3, and V_2O_5. With the exception of GeO_2, these oxides do not readily form glasses by themselves unless very rapidly quenched or vapor deposited, but can serve as glassformers when mixed with other oxides. The elements S, Se, and Te act as glassformers in chalcogenide glasses. Although halide glasses can be made in many systems, with many different compounds acting as glassformers, the two most common halide glassformers are BeF_2 and ZrF_4.

Although the number of possible glass compositions is effectively unlimited, the vast bulk of commercial glasses are based on silica as the glassformer. While silica itself forms an excellent glass, with a wide range of applications, the use of pure silica glass for bottles, windows, and other bulk commercial applications would be prohibitively expensive due to the high melting temperature (>2000 °C) required to produce vitreous silica. Production of silicate glasses requires the addition of a *flux* to reduce the processing temperature to within practical limits, *e.g.*, <1600 °C. The most common fluxes are the alkali oxides, especially Na_2O (soda), and PbO. Most commercial glasses contain soda, including those used for containers and window glasses. Potassium oxide is also used extensively in commercial glasses, while lithium oxide is used in a number of commercial glass-ceramics. Rubidium and cesium oxides are frequently used in laboratory studies of trends in behavior due to changes in the identity of the alkali oxide present in glasses, but are very rarely used in commercial products due to their high cost. PbO, which is an excellent flux, is becoming much more limited in use due to concerns regarding the toxicity of heavy metals. PbO is especially useful in dissolving any refractory or other impurity particles which might otherwise result in flaws in the final glass.

While the addition of fluxes to silica leads to decreased cost of glass formation, the addition of large amounts of alkali oxides results in serious degradation in many properties. In particular, the chemical durability of silicate glasses containing large concentrations of alkali oxides is degraded to the point where they can no longer be used for containers, windows, or insulation fibers. The degradation in properties is usually countered by addition of *property modifiers*, which include the alkaline earth and transition metal oxides, and, most importantly, aluminum oxide (alumina). While these oxides partially counter the reduction in processing temperature obtained by the addition of fluxes, they also improve many of the properties of the resulting glasses. The properties are thus modified, or adjusted, by careful control of the amount and concentration of these oxides to obtain precisely the desired results. Since many of these oxides are actually very weak fluxes for silica, and the property modifiers are usually added in lesser quantities than the fluxes, their use does not lead to excessively high processing temperatures.

Colorants are used to control the color of the final glass. In most cases, colorants are oxides of either the 3d transition metals or the 4f rare earths. Uranium oxides were once used as colorants, but their radioactivity obviously reduces their desirability for most applications. Gold and silver are also used to produce colors by formation of colloids in glasses. Colorants are only used if control of the color of the glass is desired, and are usually present in small quantities. Iron oxides, which are common impurities in the sands used to produce commercial silicate glasses, act as unintentional colorants in many products. When colorants are used to counteract the effect of other colorants to produce a slightly gray glass, they are referred to as *decolorants*.

Finally, *fining agents* are added to glassforming batches to promote the removal of bubbles from the melt. Fining agents include the arsenic and antimony oxides, potassium and sodium nitrates, NaCl, fluorides such as CaF_2, NaF, and Na_3AlF_6, and a number of sulfates. These materials are usually present in very small quantities (<1 wt %) and are usually treated as if they have only minor effects on the properties of the final glasses. Their presence, however, is essential in many commercial glasses, which would be prohibitively expensive to produce without the aid of fining agents in reducing the content of unwanted bubbles in the final product.

COMPOSITIONAL NOMENCLATURE

The terminology used to describe the composition of a given glass can be quite confusing. No single system exists for designating the composition of inorganic glasses. Not only are different systems used for oxide, halide, and chalcogenide glasses, but a number of different systems are used for oxide glasses alone. Compositions may be expressed in terms of molar, weight, or atomic fractions or percentages, depending upon the background and inclination of the authors of individual papers.

Historically, oxide glass compositions were expressed in terms of weight percentages of the oxide components, in what is known as *oxide formulations*. A composition for a *soda–lime–silicate* glass might thus be given as 15% soda, 10% lime, and 75% silica. The reader is assumed to know that the percentages are based on weights of each component and that soda is Na_2O and lime is CaO. While use of a weight fraction basis simplifies preparation of batches and is very useful in commercial production, it does little to aid in understanding the relative effect of various components on glass or melt properties. On the other hand, use of a molar fraction or percentage basis, while very useful in understanding compositional effects and currently used in most of the literature, complicates batch preparation. In either case, oxide formulations suggest that the components of the glass somehow exist as distinct, separate oxides in the melt or glass, which is certainly not the case.

Atomic fraction, or stoichiometry, formulae are routinely used to express the composition of chalcogenide glasses and for certain simple oxide compositions. A glass containing 40 atom % arsenic and 60 atom % sulfur, for example, might be designated as either As_2S_3, $As_{40}S_{60}$, or $As_{0.4}S_{0.6}$. Since the chalcogenide glasses are usually based on elements rather than compounds, this terminology works quite well in these systems. Batch calculations still require conversion to a weight basis, but the conversion is rather simple for elements. The greatest danger in this approach lies in the erroneous implication that compounds such as As_2S_3 actually exist in the glass as structural entities.

Use of the stoichiometry approach for oxide glasses, however, can be much more confusing. Consider, for example, the general formula $x\text{Li}_2\text{O}–(100-x)\text{SiO}_2$, where glasses can be made with x having any value between 0 and 40. If we use the atom % approach, this series of glasses would be described by the general formula $\text{Li}_{2x}\text{Si}_{(100-x)}\text{O}_{(200-x)}$. Certain specific compositions, *e.g.*, $x = 33.33$, can be expressed as either $33.33\text{Li}_2\text{O}–66.67\text{SiO}_2$ or $\text{Li}_{66.7}\text{Si}_{66.7}\text{O}_{166.7}$ or, with reduction to simplified forms, as either $\text{Li}_2\text{O}–2\text{SiO}_2$ or $\text{Li}_2\text{Si}_2\text{O}_5$. Since this glass contains a ratio of 1 mole of lithium oxide to 2 moles of silica, it is tempting to refer to the composition as that of a *lithium disilicate* glass, implying that the structure contains molecular units of that composition or that the structure of the glass must be based on the structure of that crystalline phase. Although all of these glasses have exactly the same composition, the use of so many different systems for compositional designation serves to confuse thoroughly the novice (and frequently even the expert!). The oxide formulation based on mol % of each oxide component will be used for oxide glasses throughout the remainder of this text in order to minimize this confusion and in keeping with the most commonly used system at present.

The terminology commonly used in the literature for halide glasses is even less consistent than that used for oxide glasses. Heavy metal fluoride glasses based on ZrF_4, for example, are usually designated by an acronym which list the glassformer first instead of last, as is usually the case for oxide glasses. The remaining components are listed in order of ascending valence of the cations, with the exception that monovalent cations are listed last. To further complicate matters, the elements are not designated by their normal chemical symbols, but rather by a set of symbols used exclusively by those working in this field. A glass containing the fluorides of Zr, Ba, Al, La, and Na would thus be designated as a ZBLAN glass, using the first letters of each component fluoride and listing the components in the order discussed above. Since the letter L has been used for lanthanum, a similar glass containing lithium instead of sodium would be designated as a ZBLALi glass.

The system used to designate the composition for halide glasses is made even more complex by a lack of consistency among various workers. Glasses in the system containing Cd, Li, Al, and Pb, which should be designated CLiAP glasses on the basis of the heavy metal fluoride nomenclature system, have been designated as CLAP glasses. Since the novice does not know that L may designate lanthanum in one glass and Li in another, the system becomes so filled with jargon that it is incomprehensible to the outsider.

BATCH CALCULATIONS

Glass batch calculations can range from very simple to very complex as a function of the complexity of the composition and the raw materials used to prepare the mixture. Batches containing only oxides in their exact state as expressed by the glass formula, for example, involve very simple calculations, while batches using a number of different minerals, where a glass component may be present in two or more raw materials, require much more complicated calculations.

All batch calculations follow the same procedure. First, determine the weight fraction of each component required to produce the desired molar composition. Begin by multiplying the mole fraction of each component by the molecular weight of that component. Next, total these contributions to determine the molecular weight of the glass, and then divide each individual contribution by the molecular weight of the glass to determine the weight fraction of each component. Finally, multiply the weight fraction of each component by the amount of glass to be produced (Example 3.1). The batch weight of any components which decompose during melting are adjusted by multiplying the weight fraction of that component by the appropriate gravimetric factor for the raw material actually used in the batch (Example 3.2). Use of raw materials which supply more than one batch component require additional calculations, as illustrated in Example 3.3.

Small compositional changes are often excluded from batch calculations. For example, it is common to use a base glass of fixed composition to study the effects of minor additions of other components. These

Example 3.1

Glass composition: $65CaO–35Al_2O_3$
Molecular weights of components (in g mol^{-1}):
$CaO = 56.08$ $Al_2O_3 = 101.96$
Molecular wt of glass: $(0.65 \times 56.08) + (0.35 \times 101.96) =$
72.14 g mol^{-1}
Weight fraction of each component:
$CaO = (0.65 \times 56.08) \div 72.14 = 0.505$
$Al_2O_3 = (0.35 \times 101.96) \div 72.14 = 0.495$
For 100 grams of glass: $CaO = 0.505 \times 100 = 50.5$ g
$Al_2O_3 = 0.495 \times 100 = 49.5$ g

Example 3.2

Glass composition: $20Na_2O–80SiO_2$
Molecular weights of components (in g mol^{-1}):
$Na_2O = 61.98$ $SiO_2 = 60.09$
Molecular wt of glass: $(0.20 \times 61.98) + (0.80 \times 60.09) =$
60.47 g mol^{-1}
Weight fraction of each component:
$Na_2O = (0.20 \times 61.98)\ 60.47 = 0.205$
$SiO_2 = (0.80 \times 60.09)\ 60.47 = 0.795$
For 100 grams of glass: $Na_2O = 0.205 \times 100 = 50.5$ g
$SiO_2 = 0.795 \times 100 = 79.5$ g
Sodium oxide is not stable in air, so we must use a batch component such as Na_2CO_3, which yields Na_2O after decomposition. It is necessary to multiply the desired quantity of Na_2O by the gravimetric factor for Na_2CO_3 (1.71) to obtain the weight of Na_2CO_3 (86.36 g) to be used to yield the desired 50.5 g of Na_2O

additions are usually expressed as wt % additions. One might, for example, consider the effect of arsenic oxide as a fining agent by preparing the same batch containing 0, 0.1, 0.2, or 0.5 wt % additions of As_2O_5 to a base soda–lime–silicate glass composition. In reality, the actual molar composition of the glass changes as the arsenic oxide content changes. If the additions are quite small, however, the base composition is only very slightly affected by these additions and the exact composition of each glass is usually not stated. While this procedure is pragmatically acceptable, one should be careful when the effects of the

Example 3.3

Glass composition: $20Na_2O$–$5Al_2O_3$–$75SiO_2$
Molecular weights of components (in g mol^{-1}):

$Na_2O = 61.98$ $Al_2O_3 = 101.96$ $SiO_2 = 60.09$

Molecular wt of glass:
$(0.20 \times 61.98) + (0.05 \times 101.96) + (0.75 \times 60.09) = 62.56$ g mol^{-1}
Weight fraction of each component:

$Na_2O = (0.20 \times 61.98) \div 62.56 = 0.198$
$Al_2O_3 = (0.05 \times 101.96) \div 62.56 = 0.0815$
$SiO_2 = (0.75 \times 60.09) \div 62.56 = 0.720$

For 100 grams of glass: $Na_2O = 0.198 \times 100 = 19.8$ g
$Al_2O_3 = 0.0815 \times 100 = 8.15$ g
$SiO_2 = 0.720 \times 100 = 72.0$ g

If we use albite feldspar as the source of alumina, we also obtain some of the soda and silica needed for the batch. Using the gravimetric factors for albite in Table 3.1, we find that 41.89 g of albite will yield the required 5.68 g of alumina. This amount of albite also yields 4.95 g of soda and 28.89 g of silica (divide the weight of albite by the gravimetric factor to find the yield for a given amount of albite). After subtracting these quantities from the required amounts of soda and sand, we find that we must add 14.85 g of soda and 43.11 g of sand. If we use Na_2CO_3 as the source of the additional soda, we will require $14.85 \times 1.71 = 25.39$ g of Na_2CO_3.

Final batch: $Na_2CO_3 = 25.39$ g
Albite = 41.89 g
Sand = 43.11 g

additions on the overall composition become too large to be ignored. In some cases, additions of 5–10 wt % of a component have been treated as negligible changes in the base composition. This is not an acceptable procedure.

MECHANISMS OF BATCH MELTING

A large number of steps occur during the conversion of a batch to a melt. While the details will be strongly dependent upon the specific batch materials used and the type of glass produced, a number of these steps occur for most batches. Since silicate glasses are most widely used commercially, far more information is available regarding these materials than for other compositions. The steps which occur during the formation of a soda–lime–silicate melt will be used as an example of the processes which may occur during melt formation.

Release of Gases

Initial heating of a glassforming batch usually results in the release of some moisture, which may have been absorbed on the particles or combined as water of hydration or as hydroxyl. Many of the components of common glassforming batch are somewhat hygroscopic, readily absorbing some water from the surroundings. Boric oxide may partially convert to boric acid (H_3BO_3), CaO may form $Ca(OH)_2$, *etc.* Other components already contain water, *e.g.*, NaOH, clays, hydrated alumina, $NaB_4O_7 \cdot 10H_2O$, which will be released at moderate temperatures. The temperature at which this water is released will depend upon the nature of its bonding to the materials, *i.e.*, physical or chemical, and the strength of these bonds. Removal of this water carries heat from the batch and increases the cost of processing.

Far more gas is released during the decomposition of carbonates, sulfates, and nitrates. The gases released expand to volumes much greater than that of the starting batch, resulting in considerable mixing and stirring action, which aids in homogenization of the melt. The creation of so much gas, however, also leads to the formation of an extremely large number of bubbles, which must be removed from the melt before processing is completed. If we consider limestone (calcium carbonate), with a density of $\approx$ 2.7 g cm^{-3}, we find that decomposition of one mole of limestone, which has a volume of 37 cm^3, produces 22 400 cm^3 of CO_2 gas, *i.e.*, an expansion in volume of a factor of 600. Since a single 1 mm diameter bubble per m^3 of glass is often considered to be a serious flaw concentration in production of commercial glasses, we find that the gas concentration of the melt must be reduced by 10–12 orders of magnitude relative to that of the starting batch!

The rapid formation of a liquid can entrap a portion of the air which initially occupies the space between particles and result in bubble formation. Rapid heating of such a melt can lead to expansion of these bubbles and foaming of the melt. It is possible for the foaming due to trapped air to cause the melt to rise above the upper rim of a crucible if the crucible is initially filled to the top with batch. Since glassforming melts are usually very efficient at dissolving the bottom of furnaces, care should always be taken to ensure that this does not occur.

Formation of Liquid Phases

Liquid phases are formed by the direct melting of batch components, by melting of decomposition products, and by melting of eutectic mixtures formed from the batch components. In soda–lime–silicate batches, we find eutectic mixtures of sodium and calcium carbonates, which melt at 775 °C, and sodium disilicate and silica, which melt at ≈800 °C. At this point, the liquid phases are very fluid, the release of gases occurs rapidly, and the mixture of liquid and solids is very turbulent. As the temperature increases, the rate of dissolution of the refractory particles such as sand, alumina, and feldspars increases. The increase in concentration of these components causes a rapid increase in viscosity and the release of additional gases as the solubility of CO_2 and other gases decreases with increasing silica concentration in the melt. Since the viscosity increases rapidly as the silica content of the liquid increases, the temperature must be increased even further to keep the melt fluid enough for thorough mixing between liquid and remaining solids. The final stage of the melting process, in which the remaining silica and other refractory components are completely dissolved and the melt becomes homogeneous, occurs much more slowly due to the high viscosity of the melt.

The time required to dissolve completely the original batch is known as the *batch-free time*. Although the definition of the batch-free time is straightforward, determination of the exact time at which the last remaining trace of batch remains in the melt is difficult. In general, determination of the batch-free time is subject to a variety of errors, including prejudice on the part of the researcher.

Studies of the batch-free time have, not unexpectedly, shown the great importance of temperature on the rate of melt formation. Other factors include overall glass composition, specific batch components used to obtain that composition, batch homogeneity, grain size of batch components, and the grain size and amount of *cullet* added to the batch. The use of cullet, or scrap glass, not only reduces waste, but also aids in reducing batch-free time both by reducing the amount of refractory material in the batch and by providing additional liquid throughout the melting process.

The overall glass composition is by far the most important factor in controlling the batch-free time. Simple oxide mixtures, such as those used to produce calcium aluminate glasses, often form eutectic mixtures which melt directly with very short batch-free times. Many non-silicate melts are very fluid at any temperature above the melting point of their components and rapidly dissolve all batch particles. Borate, phosphate, and germanate melts can be formed at much lower temperatures than are typically required for silicate melts. As a result, it is usually easier to decrease their viscosity by increases in temperature, *e.g.*, an increase in temperature from 1000 to 1200 °C is more easily attained than an increase from 1400 to 1600 °C.

The choice of batch components is also important in controlling the batch-free time. Many batch components can be supplied from a variety of raw materials. Since the different sources of a given component may have very different melting temperatures, the production of the initial liquid is highly dependent upon the choice of raw materials. Batch segregation inhibits melting by separating components which might otherwise form eutectic mixtures. Time invested in the initial mixing of batch components is almost always repaid in improved glass quality and reduced requirements for melting temperature and time.

Changes in particle size can seriously affect the batch-free time for melts. While fine particles melt more rapidly, they can also agglomerate to form larger, porous particles, which effectively prevent penetration of the viscous liquid to the particle surfaces. Since these agglomerations have a low bulk density, they can float to the surface of the melt, which significantly slows the dissolution process. Escape of gases is inhibited when very fine particles are used, since the channels between the particles are reduced in size. Use of very fine particles can result in the blockage of these channels in the early stages of the melting process, which can suppress decomposition reactions.

Melting Accelerants

A number of methods can be used to decrease the batch-free time. The most important methods for accelerating the melting process are based on changes in batch raw materials. Replacement of a small portion of sodium carbonate by sodium sulfate, for example, speeds the dissolution of sand by forming additional, lower melting eutectic mixtures. As more silica dissolves into the melt, sulfates separate from the melt in the silica-rich regions surrounding the silica particles. These sulfates decompose as the temperature increases, reacting with the silica particles to release sodium and form sodium metasilicate, which melts readily to form a fluid liquid. The release of SO_3 creates a vigorous stirring effect which aids in homogenization of the melt and improves the contact between silica particles and the surrounding liquid.

Other melting accelerants are also based on replacement of some of the sodium carbonate by more easily melted compounds such as NaOH, NaF, or NaCl, all of which form very fluid liquids upon melting. When melts are formed under an atmosphere containing oxygen, halides will gradually exchange with the oxygen in the surrounding atmosphere, the melt composition will revert to the desired oxide composition, and the viscosity will increase. Since the batch has dissolved by this time, however, the use of the halide will still result in a shorter batch-free time.

Components which force water to enter the melt are especially effective in accelerating melting processes. Water is very efficient in reducing the viscosity of oxide melts. The source of the water is unimportant, so long as it is in direct contact with the liquid. Water can be provided by use of components such as H_3BO_3 instead of B_2O_3, or NaOH instead of Na_2CO_3, by wetting the initial batch, or by changes in the furnace atmosphere which increase the partial pressure of water vapor in the combustion gases. Changes from gas–air combustion to gas–oxygen combustion result in higher water vapor contents in furnace atmospheres, which may lead to reductions in overall melting times.

Mechanical methods which improve contact between batch particles may also accelerate melting processes. Compaction of raw materials into either dry or moist bricks, pellets, or granules increases the melting rate by improving heat transfer to the individual particles and by providing better contact between components which will react to form eutectic melts.

Volatilization of Components from Melts

A large number of the components of glasses are quite volatile at elevated temperatures. Loss of these components can significantly alter the composition of the glass obtained after prolonged melting compared to that obtained for short melting times. Volatilization losses are particularly significant for alkali oxides, lead, boron, phosphorus, halides, and other components which have high vapor pressures at high temperatures. The rate of loss of alkali increases rapidly in the order of $Li < Na < K < Rb < Cs$. The loss of a component can be reduced by increasing the concentration of that component in the atmosphere above the melt. Covering the melt will force the partial pressure of the volatile components to increase directly above the melt, establishing a dynamic equilibrium between the dissolved and vaporized species, and prevent significant loss of those components. These losses can usually be reduced dramatically by lowering the melt temperature. Covering the melt is usually not possible in large commercial melting tanks. It may be necessary to allow for losses by providing excess concentrations of components known to vaporize from a given melt. This procedure is quite efficient for continuous melting of a constant composition, where analysis of the product over a period of time can be used to establish the exact amount of excess component needed to counter the volatilization losses. In a few cases, it may be possible to add a component to the furnace atmosphere in order to prevent its loss from the melt. This procedure can be used to aid in retaining halides or water in melts, but is of little use for cations.

While control of volatilization losses is a well-recognized problem in the production of commercial glasses, it is often totally overlooked in laboratory studies. Many papers deal with systems where volatilization from the melt is an obvious source of potential error in glass compositions. Since laboratory melts are often held at times and temperatures far in excess of those actually needed to produce a bubble-free, homogeneous glass, volatilization of components can seriously alter the composition of the resulting glass. Unfortunately, very few such studies make any attempt to determine the possible changes in composition which may occur during melting. Although it would be best if every glass discussed in scientific papers were analyzed for all components, this is almost never done. The lack of analysis is often due to the expense and time needed for such analysis or due to the difficulty encountered in analysis of certain components such as fluorine.

Although a complete chemical analysis is rarely practical, there is a simple method which, if widely used, would detect many errors due to volatilization losses. A simple weight loss method can be used to determine if the yield of glass is equal to that predicted in the batch calculation. If the yield differs from the expected value by more than a few tenths of a percent, the sample composition is certainly suspect. Experience has shown that production of a soda–lime–silicate glass using Na_2CO_3, CaO, and SiO_2 will result in a loss of $\approx$0.1 wt % relative to the expected yield. This loss is usually due to adsorbed water vapor on the batch materials and not to detectable loss of sodium. This loss can be contrasted with values of 10–20 wt % occasionally observed for melts containing rubidium, cesium, thallium, and other highly volatile, high atomic weight species. While this method does not reveal the identity of the species lost, it certainly alerts the researcher to the possibility of serious error in the sample composition.

FINING OF MELTS

The terms *fining* and *refining* refer to the removal of gaseous inclusions, or *bubbles*, from the melt. Although the presence of bubbles in a glass sample is not necessarily detrimental for many scientific studies, bubbles are definitely undesirable in most commercial glasses. Bubbles in commercial products are almost always considered flaws and result in rejection of the product. These gas-filled inclusions may occur as very small spheres (<0.4 mm diameter), which are frequently called *seed* and are often found in clusters, or as larger, isolated spheres more commonly referred to as bubbles. Fining of a melt begins during the melting process, but typically extends to times well after the complete disappearance of residual batch.

Sources of Bubbles

Bubbles can be formed by the physical trapping of atmospheric gases during the initial phase of batch melting or by the decomposition of batch components. The gases in the interstices between batch particles may be trapped as the particles begin to soften and form a viscous liquid. As the viscosity decreases with increasing temperature, these interstices become fully surrounded by liquid. Surface forces then cause these interstices to assume the spherical shape of bubbles. These bubbles contain gases characteristic of the melting atmosphere, which may be air, combustion gases, or some gas deliberately introduced to control chemical reactions with the batch. Prevention of the formation of these bubbles involves elimination of gas from the interstices of the batch, *i.e.*, melting under vacuum.

Trapping of atmospheric gases is enhanced by use of very fine sand in the batch or by the use of batch components of widely differing particle sizes. Both factors create a large concentration of very fine interstices within the unmelted batch. Agglomeration of particles is particularly effective in trapping atmospheric gases within the agglomerate, which is effectively surrounded by the viscous melt. Mechanical stirring of a melt can introduce bubbles by trapping air and forcing it into the melt.

Decomposition of batch materials can produce extremely large quantities of gases such as CO_2, SO_3, NO_x, H_2O, *etc.* Reactions with metals in contact with the melt can generate oxygen, carbon dioxide, or hydrogen by electrolytic reactions. Corrosion of refractories can open previously closed pores to the melt, releasing the gas contained in those pores into the melt. Residual carbon in refractories, or carbide refractories such as SiC, can react with oxide melts to form CO_2 or CO. The products of all of these reactions can agglomerate to form bubbles.

Bubbles can also be formed by precipitation from the melt whenever *supersaturation* occurs for a specific gas. Since many gases have a large enthalpy of solution in glassforming melts, their solubility in these melts is a strong function of temperature. Species which alter their chemical form with temperature or changes in melt composition are particularly susceptible to precipitation from melts where they were previously soluble. Carbon dioxide, for example, is present in silica-rich melts as CO_2 molecules, whereas it chemically reacts with alkali-rich melts to form carbonate species, which are far more soluble than the molecular species. The solubility of carbon dioxide in the alkali-rich regions surrounding silica grains during the early stages of batch melting is quite high. As the silica grains dissolve, the melt becomes locally enriched in silica, which converts the dissolved species to CO_2 molecules, which have a much lower solubility. The melt thus become supersaturated locally and CO_2 bubbles form. As a result, CO_2 bubble generation will continue long after the initial decomposition of the carbonate batch components.

Sulfur solubility is very sensitive to the oxidation state of melts. Under reducing conditions, sulfur dissolves as sulfide ions, while oxidizing conditions produce sulfate ions. Reduction of the oxygen partial pressure over an oxidized melt will shift the equilibrium between sulfide and sulfate states, with a reduction in overall sulfur solubility. A previously undersaturated melts will now become supersaturated and bubbles will form. Changes in melt chemistry will alter sulfur solubility just as they alter carbon dioxide solubility, with similar results. Melt temperature is also very important, since sulfate solubility decreases with increasing temperature, while sulfide solubility increases with increasing temperature. Changes in temperature can thus alter the degree of solubility of sulfur in melts in opposite directions for oxidized and reduced melts.

Oxygen can also be released into melts by changes in the oxidation state of the surrounding atmosphere. Changes in oxidation state of polyvalent ions such as iron, chromium, manganese, *etc.*, can alter the state of oxygen from chemically bound to physically dissolved molecules, as in the reaction

$$4Fe^{3+} + 2O^{2-} \rightarrow 4Fe^{2+} + O_2 \uparrow \qquad (3.1)$$

Since the solubility of molecular oxygen is much less than that of chemically bound oxygen, supersaturation occurs and oxygen bubbles form during the reduction of polyvalent species.

Reboil, or the formation of bubbles in previously bubble-free glasses, is a special case of the precipitation phenomena discussed above. Reboil specifically refers to bubble formation from bubble-free materials during reheating of glasses from the solid state, or during increases in melt temperature. Gases which increase in solubility with increasing temperature do not contribute to reboil, since supersaturation occurs because the solubility of the dissolved gas is less at the higher temperature. Reboil is especially common for SO_3, where the solubility in commercial soda–

lime–silicate melts decreases by three orders of magnitude on heating from 1100 to 1400 °C. Under these conditions, a small increase in temperature can convert a previously undersaturated melt into a highly supersaturated one, with the consequent formation of many bubbles. Sulfur-induced reboil can also occur if previously oxidized melts containing sulfates are placed in contact with previously reduced melts containing sulfides. The change in melt chemistry at the interface can result in a supersaturated solution due to the differing solubilities of the two forms of dissolved sulfur.

Removal of Bubbles by Buoyancy Effects

Bubbles can be removed from melts either by physically rising to the surface or by chemical dissolution of the gas into the surrounding melt. Since the density of a bubble is less than that of the surrounding liquid, a bubble will automatically rise to the surface and burst unless prevented from doing so by some external agent. The basic principle of buoyancy is given by Stokes' law, which applies to the velocity, V_s, of a solid sphere in a liquid of a different density

$$V_s = \frac{2g\Delta\rho r^2}{9\eta} \tag{3.2}$$

where g is the gravitational acceleration, $\Delta\rho$ is the difference in density between the sphere and the liquid, r is the radius of the sphere, and η is the viscosity of the liquid. It has been shown that the special case of a gas bubble in a viscous liquid is actually better represented by the slightly different expression

$$V_b = \frac{3}{2} V_s = \frac{g\Delta\rho r^2}{3\eta} \tag{3.3}$$

where V_b is the rate of rise of the bubble. In fact, the theoretical behavior of a bubble actually varies between the limits given by V_s and V_b as the size of the bubble increases. While this difference in velocity appears small, it can result in a 50% increase in fining time for small bubbles. Experimental evidence suggests that Equation 3.3 applies to the rise of $\approx$1 mm diameter bubbles in unstirred oxide melts.

Equations 3.2 and 3.3 indicate that the velocity of rise of a bubble is inversely proportional to the viscosity of a melt, *i.e.*, that bubbles will rise faster in a more fluid melt, and directly proportional to the density of the melt, *i.e.*, bubbles rise more rapidly in a more dense melt than in a less dense one. Although these predictions are generally true, there is some evidence that melt chemistry can alter this simple relationship, so that these equations do not perfectly predict the results of all experimental studies.

These equations also predict that the rate of removal of bubbles will be proportional to the square of the bubble radius or diameter. Procedures which increase bubble size will therefore rapidly accelerate fining. Conversely, these equations indicate that bubble rise is not a very efficient process for the removal of very small bubbles. It follows that very fine seed cannot be efficiently removed from viscous melts by simple bubble rise in the absence of any other processes.

If bubbles do not rise sufficiently fast in a quiescent melt, the fluid itself can sometimes be moved by convection or stirring in such a manner that the bubbles are carried to the surface. Upward fluid motion can be obtained by mechanical stirring, by design of a glass tank floor to produce upward currents, by localized heating to produce a locally hotter and thus less dense region in the melt, or by bubbling with a gas introduced near the bottom of the melt. In practice, all of these techniques have been used to varying degrees in either tank or crucible melting.

Fining Agents

Chemical methods for removal of bubbles from melts depend upon the addition of batch components collectively termed *fining agents*. Fining agents release large quantities of gases, which form large bubbles that rapidly rise to the surface of the melt. These large bubbles tend to carry smaller bubbles and seed to the surface as well. In addition, some fining agents cause the absorption of O_2 from bubbles at lower temperatures, thus reducing the size of seed due to diffusion from the bubble into the melt. The seed eventually shrink to below the critical radius, where the surface energy causes the complete disappearance of the bubble.

Arsenic and antimony oxides are the most efficient and thoroughly studied chemical fining agents. While the details of the mechanisms by which they act to remove bubbles are still controversial, no one questions their usefulness as fining agents, especially when combined with alkali nitrates in the batch. Arsenic and antimony oxides are usually added to the batch in 0.1–1 wt % quantities as trioxides, but pentoxides, arsenates, antimonates, and arsenic acid are also used under special circumstances. Alkali and alkaline earth oxides in the melt allow the formation of arsenates, which are less volatile than the trioxides. The absence of these basic oxides can seriously compromise the efficiency of arsenic and antimony oxides as fining agents owing to the rapid loss of the fining agent through volatilization.

The fining action of arsenic and antimony oxides results from a series of chemical reactions which occur at different stages of the melting process. Although controversial, one possible sequence of events might be described as follows. During batch melting, these oxides react with nitrates to release nitrogen oxides and O_2, as, for example, in the reaction

$$4KNO_3 + 2As_2O_3 \rightarrow 2K_2O + 2As_2O_5 + 4NO \uparrow + O_2 \uparrow \quad (3.4)$$

Analogous reactions occur for Sb_2O_3 and for other nitrates. These reactions release copious quantities of NO and O_2, which form large bubbles that rapidly rise to the surface. These bubbles stir the batch and sweep smaller bubbles to the surface.

After batch decomposition is completed, melts are usually heated to higher temperatures and held until completely fined. Since the trioxide is more stable than the pentoxide at these temperatures, the pentoxide produced *via* reaction with nitrates decomposes *via* the reaction

$$As_2O_5 \rightarrow As_2O_3 + O_2 \uparrow \quad (3.5)$$

for arsenic and the analogous reaction for antimony. These reactions produce O_2 bubbles, which can either form new bubbles or diffuse into nearby smaller bubbles, thus increasing their size and rise rate. Any bubbles remaining in the melt at this point will be highly enriched in oxygen relative to the bubbles previously present.

The equilibrium expressed by Equation 3.5 is very temperature dependent. Lowering the temperature will result in a shift toward the pentoxide, which requires absorption of oxygen from the melt. As the dissolved oxygen is consumed, oxygen from nearby bubbles will diffuse into the melt, reducing their internal pressure and hence reducing their diameter. Once the diameter is reduced to a very small value (<0.1 mm), the surface energy will become sufficient to cause contraction of the bubble and a rise in internal pressure, P, as described by the expression

$$P = \frac{2\gamma}{r} \quad (3.6)$$

where γ is the surface energy and r is the bubble radius. The high internal pressure will cause accelerated diffusion from the bubble into the melt, until the bubble is completely consumed.

Although other fining agents are usually less efficient, the toxicity of arsenic and antimony oxides often force the use of other approaches to chemical fining. Sodium sulfate, for example, also serves as a source of considerable gas during batch decomposition, as well as providing a portion of the sodium for soda–lime–silicate melts. The extreme temperature and compositional sensitivity of sulfur solubility in melts plays an important role in the fining process. Sulfates readily dissolve into the alkali-rich regions of melts during batch decomposition. As the silica dissolves into these alkali-rich regions, the solubility of sulfates decreases, releasing gas into the melt, as described by the reaction

$$Na_2SO_4 + nSiO_2 \rightarrow Na_2O{-}nSiO_2 + SO_3 \quad (3.7)$$

The SO_3 immediately decomposes into O_2 and SO_2 via the reaction

$$2SO_3 \rightarrow 2SO_2 + O_2 \quad (3.8)$$

The SO_2 molecules diffuse into nearby bubbles, while the O_2 either diffuses into bubbles or dissolves into the melt. These bubbles then rise to the melt surface. Lowering the temperature can result in reabsorption of SO_2 from bubbles to form sulfate chemically dissolved in the melt, which will cause the bubbles to shrink and disappear in a manner similar to that described for arsenic and antimony fining.

Sulfate fining is strongly affected by the reactions with furnace gases or other sources of carbon. Reactions between SO_3 and either C or CO produce SO_2 and CO_2, releasing large quantities of gas. Since carbon serves as both a reactant to produce CO_2 and as a reducing agent, which alters the solubility of sulfides and sulfates in melts, the chemistry of sulfate–carbon fining is very dependent upon the overall melt composition, temperature, and surrounding atmosphere.

Nitrates can also act as fining agents even in the absence of arsenic or antimony oxides. Decomposition of nitrates releases large quantities of nitrogen and oxygen, which form large bubbles. The low decomposition temperatures (500 to 800 °C) of alkali nitrates, however, limits their usefulness as fining agents since most of the gas release occurs before the melt forms.

Halides are most useful as fining agents through their efficiency in lowering the viscosity of melts. Since exposure of melts containing halides to oxygen atmospheres will result in a gradual replacement of the halide by oxygen, the viscosity will gradually rise. Since most of the bubbles will have already risen to the surface, this delayed viscosity increase may not necessarily be detrimental to the fining effect of halides. Halides are particularly effective in fining of high alumina content melts.

Oxides of a few polyvalent cations can be used as fining agents by acting as sources of O_2, in a manner similar to arsenic and antimony oxides. Cerium oxide, for example, can be reduced to release oxygen *via* the reaction

$$4CeO_2 \rightarrow 2Ce_2O_3 + O_2 \quad (3.9)$$

Since cerium oxide is often added in small quantities to commercial glassforming melts for other purposes, *e.g.*, prevention of solarization, the fining effect can be considered a beneficial side effect. Other oxides of polyvalent cations (MnO_2, Fe_2O_3, Pb_3O_4, *etc.*), which are added for reasons other than their fining action, can also act as minor fining agents via similar reactions.

HOMOGENIZING OF MELTS

The fluid produced during the initial batch decomposition process is very heterogeneous. This heterogeneity is gradually reduced by the stirring action of rising bubbles during the fining process. Production of an acceptably homogeneous glass, however, usually requires additional time for diffusion processes to improve the homogeneity of the melt. Homogeneity is normally described in a negative sense: a homogeneous melt is free from significant heterogeneities. Definition of a 'significant' heterogeneity is often difficult and varies dramatically with the intended end use of the glass. A perfectly acceptable level of homogeneity for a window glass might be completely unacceptable for a glass used as an optical lens.

Gross defects such as bubbles, seed, and *stones* (particles of undissolved material) are often clearly visible and result in rejection of most commercial glasses. The terms *striae* and *cord* are used to describe variation in local composition within a glass. Striae are two-dimensional regions, or layers, of a composition different from the bulk, while cord are similar regions which are effectively one-dimensional veins in the glass. These regions are most often detected by visual observation, where a 'wavy' appearance is caused by local refractive index variations. In colored glasses, regions of inhomogeneity can often be detected by variations in color intensity. The extent of these regions is designated as the *scale* of the inhomogeneity, while the degree of departure from the bulk composition, or property, is designated as the *intensity* of inhomogeneity.

Quantitative description of the degree of inhomogeneity of a glass is difficult. Density or refractive index measurement on crushed glasses can be used as a relative measurement of intensity of inhomogeneity. Visual examination of a pattern consisting of dark, parallel lines through a glass plate is frequently used as a method for detection of optical distortions due to cord and striae. Since the pattern resembles the coat of a zebra, this technique is called the *zebra board* method.

Poor homogeneity frequently results from poor mixing of the original batch materials. The effect of mixing is especially important for laboratory melts, which are usually not stirred and are held at melt temperatures for times which are much less than those used for commercial glass production. Striae and cord can be formed by reactions with refractories and at the melt–atmosphere boundary owing to volatilization of batch components, especially alkali, boron, or lead. Decreasing the grain size of the batch improves homogeneity by reducing the scale of the inhomogeneity of the initial melt. Stirring by mechanical stirrers, creation of convection flow in the melt, or bubbling of a gas through a melt can improve homogeneity.

SPECIALIZED MELTING METHODS

Many glassforming melts require special techniques which do not apply to the melting of more common compositions. Toxicity may require processing in glove boxes to protect workers from dangerous fumes or powders. Volatility of components may require processing in sealed containers to obtain the desired composition of the final glass. Special atmospheres may be needed to prevent contamination of non-oxide glasses.

Formation of chalcogenide glasses provides an excellent example of the complexity encountered in melting unusual glasses. Not only are the components of these glasses toxic and highly volatile, but contamination by very small quantities of oxygen will destroy the infrared transmission of the glass. These melts are usually prepared by weighing and mixing the components in an inert atmosphere dry box, placing the mixture of powders in a vitreous silica tube, and then sealing the tube at both ends under a vacuum. The tube is placed in a furnace, heated to the desired temperature, and then quenched to allow the melt to form a glass. Since homogeneity of the melt is difficult to attain under these conditions, the ampoules are often heated in a rocking furnace to aid in stirring of the melt.

Heavy metal halide glasses also require melting in an oxygen-free atmosphere to preserve their optical properties. These glasses are often melted under a reactive atmosphere, such as CCl_4 or SF_6, which, owing to decomposition of the atmospheric gas, contains free halide. The atmosphere acts as both a getter for oxygen and as a source of halide to replace volatilization losses and maintain the stoichiometry of the melt.

This melting procedure is termed *reactive melt processing*, usually designated as RAP.

SUMMARY

Production of glasses by melting involves four steps: batching, batch melting, fining, and homogenization. Batching involves selection of raw materials, calculation of the concentrations of each material, weighing, and mixing of powders and, occasionally, liquids. Batch melting involves the decomposition of the raw materials to form the initial melt and control of temperature and atmosphere during the time of formation of the liquid. Specialized techniques are often required for production of glasses from non-oxide and/or toxic materials. Fining, or removal of bubbles, occurs by either bubble rise or bubble absorption. Bubble rise is assisted by formation of large bubbles during batch decomposition, by low melt viscosity, and by use of proper particle sizes for batch components. Special chemicals known as fining agents are often added to batches to promote fining. Finally, production of a homogeneous glass requires elimination of the heterogeneities inherent in producing a melt from a large number of components with widely different properties.

Chapter 4

Immiscibility/Phase Separation

INTRODUCTION

If we pour a small amount of alcohol into a flask containing water, we find that a single homogeneous liquid exists over a broad compositional range of relative alcohol and water concentrations. These liquids are said to be completely *miscible.* In contrast, if we pour a small amount of oil into a similar flask containing water or vinegar, we find that a single liquid is not stable and that the mixture will spontaneously separate into two liquids, with the less dense floating on the surface of the more dense. In this case, the liquids are said to be *immiscible,* and the mixture is said to be *phase separated.*

If we seal the opening in the flask containing the oil and water and vigorously shake the mixture, we will form small, dispersed droplets of one of the liquids in a matrix of the other liquid. Since oil and water are both very fluid liquids, the more dense liquid will rapidly sink to the bottom of the flask, while the less dense liquid rises to the surface. The original two layered structure will rapidly reappear. If we carefully pour the contents into a beaker, we can recover almost all of the oil, with very little contamination by water.

What change in this behavior would we observe if these liquids were very viscous? In this case, the rate of separation would be severely depressed. If we rapidly decreased the temperature to below the freezing temperature of both liquids, we would obtain a solid which consists of solid, dispersed droplets of one material in a continuous matrix of the other. The identity of the matrix and droplet materials would be controlled by the relative concentrations of oil and water in the mixture. A mixture which originally contained mostly water would form oil droplets in a water matrix, while one which contained mostly oil would form water droplets in an oil matrix.

If we now consider the liquid formed during melting of a glass batch, we can see that this liquid might, in some cases, also spontaneously separate into two very viscous liquids, or phases. Since these liquids are initially intimately mixed, complete separation into two layers would be a very slow process. Cooling the melt to a temperature below the glass transformation region would be equivalent to freezing the oil and water mixture discussed above. The resulting glass is now said to be *phase separated* as a result of *liquid–liquid immiscibility*. Since the material is now a solid, no further separation would occur unless the glass were reheated to a temperature where flow processes would allow separation to continue.

Many glassforming melts exhibit liquid–liquid immiscibility. In some cases, the liquids are so fluid that complete separation into two layers readily occurs. If we cool a crucible containing such a melt without disturbing the layers, we can often separate the resulting sample into two pieces of glass, each of which has a different composition and properties. On the other hand, the melt may be so viscous that the degree of separation is very small, so that the droplets cannot be detected without use of very high magnification, *i.e.*, electron microscopy, and the sample appears to be a homogeneous glass. No evidence of phase separation can be detected by the naked eye. A large number of commercial sodium borosilicate glasses exhibit this type of phase separation.

THERMODYNAMIC BASIS FOR PHASE SEPARATION

Liquid–liquid immiscibility, or phase separation, is a common phenomenon in liquid systems. It should not be surprising to find that it is also a common phenomenon in melts. Actually, far more binary glassforming melts exhibit liquid–liquid immiscibility than exhibit homogeneous liquid behavior. Formation of homogeneous glasses from melts might therefore be considered to be rather unusual.

Why do some liquids or melts separate into two liquid phases while others remain homogeneous? The answer, as usual, is found in the behavior of the free energy of the system. If mixing of two components yields a lower free energy, the mixture will remain homogeneous. If, however, separation of the mixture into two components yields a lower free energy, separation will occur if allowed by kinetic considerations.

A simple model for the thermodynamic basis for phase separation can be derived from consideration of the *free energy of mixing*, ΔG_m, which is given by the expression

$$\Delta G_m = \Delta H_m - T\Delta S_m \tag{4.1}$$

where ΔH_m is the *enthalpy of mixing* and ΔS_m is the *entropy of mixing*. For the simple case of a *regular solution*, the entropy of mixing at temperature T is given by the expression

$$\Delta S_m = -R[X_1 \ln X_1 + X_2 \ln X_2] \quad (4.2)$$

where X_1 is the concentration expressed as a mole fraction of phase 1, X_2 is the concentration expressed as a mole fraction of phase 2, and R is the gas constant. The enthalpy of mixing is given by the expression

$$\Delta H_m = \alpha X_1 X_2 \quad (4.3)$$

where α is a constant related to the energies of the bonds among the various components. Combining Equation 4.1 through 4.3, we obtain the expression

$$\Delta G_m = \alpha X_1 X_2 + TR[X_1 \ln X_1 + X_2 \ln X_2] \quad (4.4)$$

If we consider the expression for the entropy of mixing, we find that, since X_1 and X_2 are less than unity, the term in brackets will always be negative (the logarithm of a number less than unity is negative). The value of ΔG_m will be either positive or negative, depending on the value of α. If α is negative, ΔG_m will always be negative, with a minimum at $X_1 = X_2$, and the system will not exhibit phase separation. If, however, α is positive, there will be a competition between the contributions from the enthalpy and entropy of mixing, and the sign on ΔG_m will be a function of temperature. At $T = 0$ K, the entropy term $T\Delta S_m$ will be zero, and the free energy will be positive. As a result, the mixture will separate into the end member components if allowed by kinetics. At a sufficiently high temperature, usually designated as the *upper consolute*, or *critical temperature*, T_c, the $T\Delta S_m$ term will always dominate and the free energy of mixing will always be negative. The system will thus be homogeneous for any temperature above T_c. Since $T_c = \alpha/2R$, the value of α can be calculated once T_c is known from experimental measurements.

At intermediate temperatures between 0 and T_c, the competition between the enthalpy and entropy terms will result in a saddle in the free energy *versus* composition curve, as is shown in Figure 4.1. This curve has two minima, which approach each other with increasing temperature, until they merge at T_c. If we draw a tangent between the minima, we find that the free energy will be less if the liquid separates into two phases, with compositions A and B, than if it remains homogeneous so that the free energy is given by the solid curve. The actual compositions of phases A and B, which are read from the abscissa of the phase diagram, will be a function of temperature, with a decreasing difference

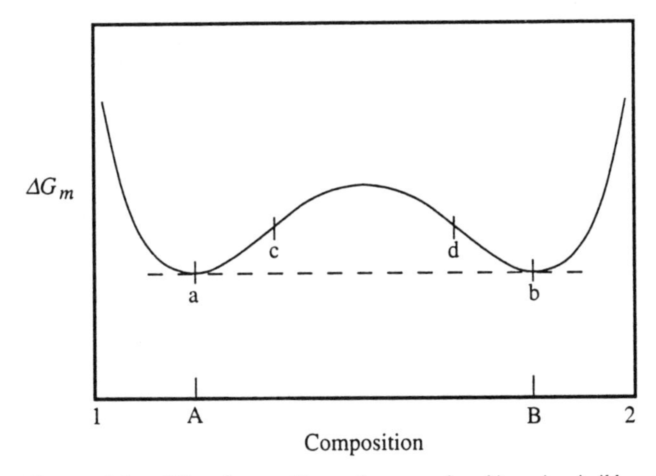

Figure 4.1 *Effect of composition on free energy for a binary immiscible system*

in composition as the temperature approaches T_c. The phase diagram resulting from this behavior will exhibit a symmetric dome, below which the liquid will spontaneously separate, if allowed by kinetics, into two components, or phases, A and B, whose compositions are given by their location on the dome. The curve describing this dome is called the *binodal, immiscibility limit* or *boundary*, or the *phase boundary*. The region inside the dome is often called the *miscibility gap* (region with a lack of miscibility) or *immiscibility region* (region where immiscibility occurs). The line connecting phases A and B is known as a *tie-line*. The tie-line will connect the compositions of the phases in equilibrium at a specified temperature. A different tie-line exists for each temperature, with different compositions for the two phases.

The relative concentrations of phases A and B can be determined by the familiar *lever rule* used in all phase diagrams. The *fractional concentration*, or relative proportion, of phase A is given by the distance between the bulk composition Y and the composition of B, divided by the distance from A to B. The fractional concentration of phase B is given by the distance between the bulk composition and the composition of A, divided by the bulk composition. Expressed in a different form

$$[A] = \frac{(B - Y)}{(B - A)} \tag{4.5}$$

and

$$[B] = \frac{(Y - A)}{(B - A)} \tag{4.6}$$

where the [ ] indicate the concentration. Since these are fractional concentrations

$$[A] + [B] = 1.0 \tag{4.7}$$

so that only the concentration of either A or B must be determined to define the concentration of both phases.

MECHANISMS FOR PHASE SEPARATION

The thermodynamic model presented above only predicts when phase separation will occur. There are, however, two mechanisms by which phase separation can actually occur. The first mechanism is similar to that discussed in Chapter 2 for precipitation of crystals from a melt, where a nucleus is formed and then grows with time. By analogy, this mechanism is termed *nucleation and growth.* Many of the same factors which control crystal formation also affect phase separation by this mechanism. The second mechanism is termed *spinodal decomposition.* This mechanism involves a gradual change in composition of the two phases until they reach the immiscibility boundary.

The mechanism which occurs is determined by the local curvature of the free energy of mixing at the bulk composition of the melt. If we examine Figure 4.1, we note that compositions lying between points a and c, and those between points d and b, where c and d are inflection points on the free energy of mixing curve, lie in regions where the second derivatives of the free energy of mixing with composition are positive. This is known as the *metastable region.* If we attempt to alter the composition of a bulk liquid lying in either region by a slight amount, *i.e.*, by separation into two liquids of similar composition, the free energy of mixing will increase and the system will tend to return to the homogeneous state. A large change in composition must occur to allow an overall decrease in free energy. If the bulk composition lies in the region a–c, for example, the composition of one phase must change to that of liquids near point b in order to reduce the free energy of mixing. Since thermodynamics require that the compositions change to the lowest energy states, the actual compositions will change to those represented by points a and b.

Bulk compositions which lie in the region between points c and d lie in a region where the second derivative of the free energy of mixing is negative. It follows that a small change in composition by separation into

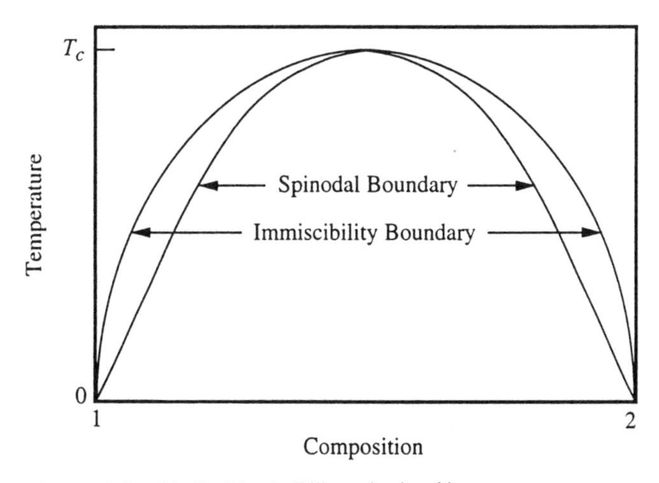

Figure 4.2 *Idealized immiscibility region in a binary system*

two phases will decrease the free energy. Since these changes occur spontaneously, this region is *unstable* with regard to immiscibility. Any fluctuation in local composition will tend to grow and the compositions of the two phases will gradually change until they reach the compositions represented by points a and b. Phase separation in this region occurs by spinodal decomposition. The curve representing the positions of the inflection points as a function of temperature is called the *spinodal boundary* and is often added to phase diagrams such as that shown in Figure 4.2 as a dashed or dotted line.

The mechanisms of nucleation and growth and spinodal decomposition result in quite different microstructures in the glass formed. Since nucleation and growth closely resembles crystallization, the microstructure formed by this process has some similarities to that found in crystallizing samples. Growth occurs on individual, isolated nuclei, so that the regions of second phase formation are clearly separated. Since the second phase is a liquid, the surface energy will be minimized for spheres, so the second phase will occur as isolated spheres of one equilibrium composition randomly dispersed through a matrix of the other equilibrium composition. The growth behavior will be such that the spheres will have the composition of the phase with the lesser volume fraction, *i.e.*, that which differs the most from the bulk composition of the melt. Since nucleation occurs randomly throughout the melt, the second phase will also occur randomly. Local connectivity of spheres may exist when two neighboring spheres intersect, but the connectivity of the minor phase will generally be quite low.

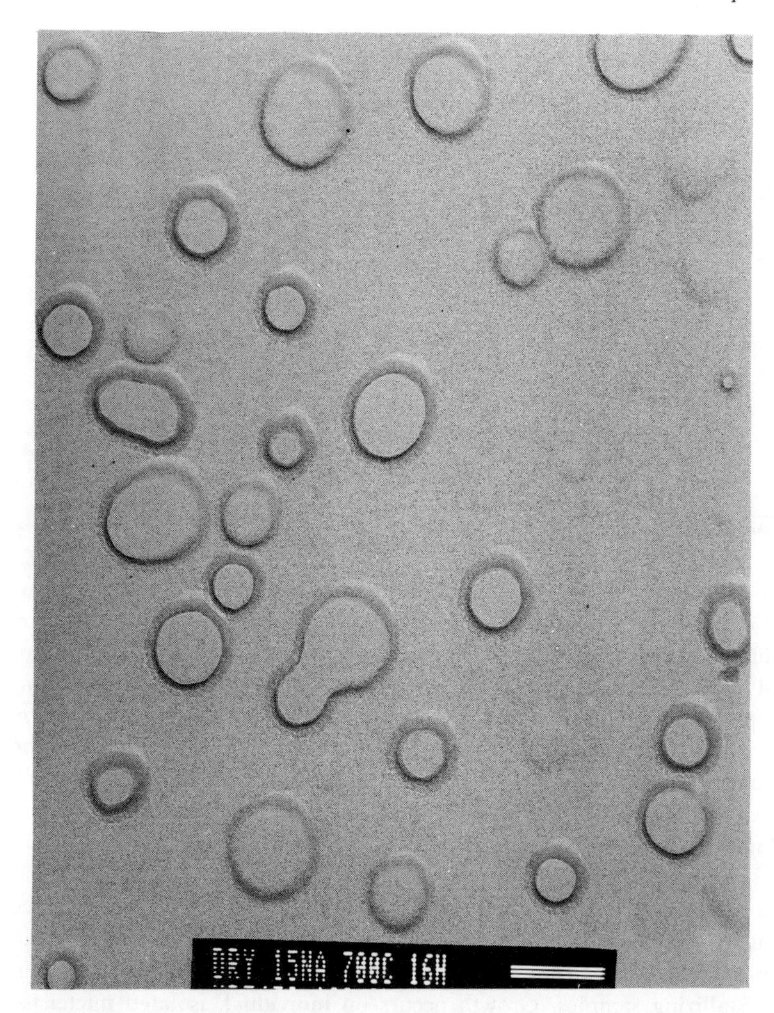

Figure 4.3 *Micrograph of a phase separated glass with a 'sphere in a matrix' morphology*

The morphology developed by spinodal decomposition will be quite different from that due to nucleation and growth. Both phases will gradually and continually change in composition until they reach the compositions of the equilibrium liquids. The interface between the phases will initially be very diffuse, but will sharpen with time. The second phase will be regularly distributed in space and characterized by a regular size. The distance between centers of either phase is sometimes termed the *wavelength* of the microstructure. Finally, and perhaps most importantly, both phases will have a high degree of connectivity, so that

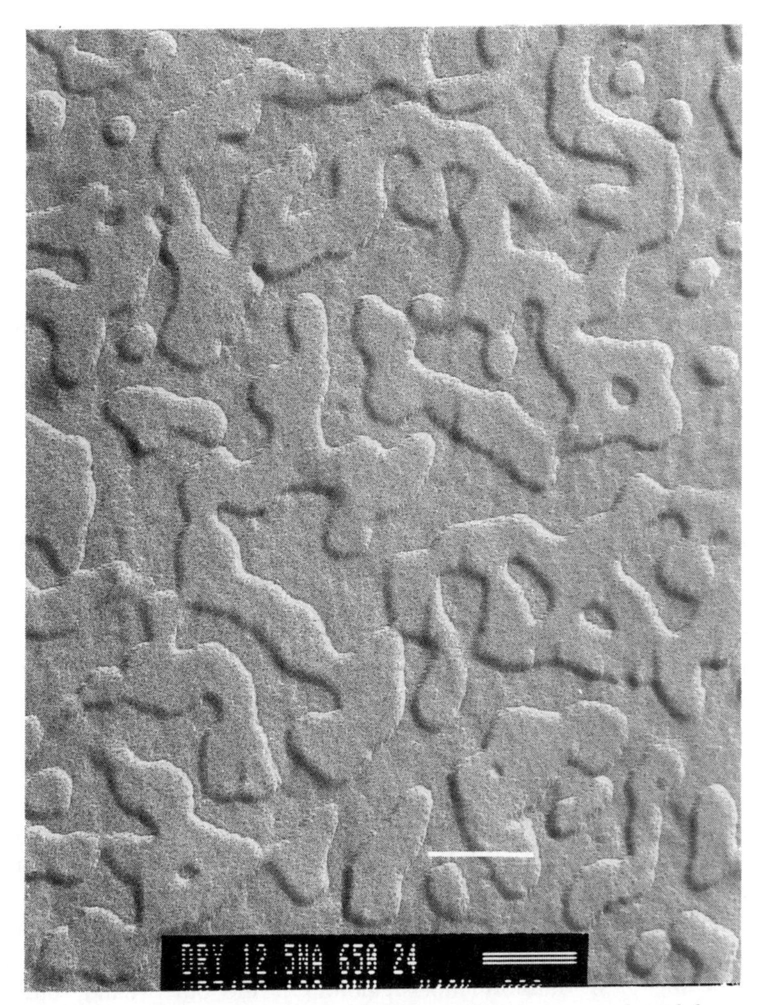

Figure 4.4 *Micrograph of a phase separated glass with an interconnected morphology*

continuous pathways through the material exist for each phase. Morphologies of this type are said to be *interconnected* and the region beneath the spinodal boundary is sometimes referred to as the *interconnected region.*

Replica micrographs taken from sodium silicate glasses exhibiting phase separation caused by either nucleation and growth or spinodal decomposition are shown in Figures 4.3 and 4.4, respectively. Since microstructures like that shown in Figure 4.3 are characterized by spheres of the minor phase dispersed in a matrix of the major phase, they are frequently called *spheres in a matrix* microstructures. The spheres are essentially isolated from one another, although an occasional area exists where two spheres have intersected during growth.

The microstructure shown in Figure 4.4 is typical of a spinodal structure. Each phase exhibits continuous pathways through the structure in three dimensions, with roughly constant path widths. The concentrations of the two phases are approximately equal in this sample, so that the areas of the micrograph covered by each phase are also approximately equal. A line drawn across this micrograph would encounter each phase with far more regularity than any similar line drawn across the micrograph shown in Figure 4.3.

IMMISCIBILITY IN GLASSFORMING SYSTEMS

So many glassforming melts exhibit liquid–liquid immiscibility that it might be argued that the occurrence of phase separation in glasses is the norm rather than the exception. The extent of phase separation can range from the sub microscopic level which requires detection by use of electron microscopes to the macroscopic level represented by complete separation into two layers in the melt. These differences in the effect of phase separation on the physical appearance of glasses are usually the result of differences in the kinetics of the separation process. Since viscosity plays a major role in determining the rate of mass transport in melts, it is not surprising to learn that viscosity has a major effect on the kinetics of phase separation.

Thermodynamic factors indicate that a system which exhibits liquid–liquid immiscibility should be phase separated at any temperature between absolute zero and T_c. Pragmatically, however, we realize that phase separation will not occur if the viscosity of the melt is too great. For most purposes, this means that no changes in morphology occur below the glass transformation region. On the other hand, phase separation will never occur in a melt held at a temperature greater than T_c. It appears, then, that a temperature window exists between T_c and $\approx T_g$ where a melt can change its microstructure through phase separation.

If we initially heat a batch to a temperature above T_c, we will obtain a homogeneous liquid. If this liquid could be instantaneously quenched to a temperature below T_g, theory then indicates that a homogeneous glass would be formed. Since an instantaneous quench would prevent any mass transport, no phase separation would occur. Of course, we cannot actually cool a melt instantaneously. As the temperature decreases, we will encounter the upper immiscibility boundary and the melt will begin

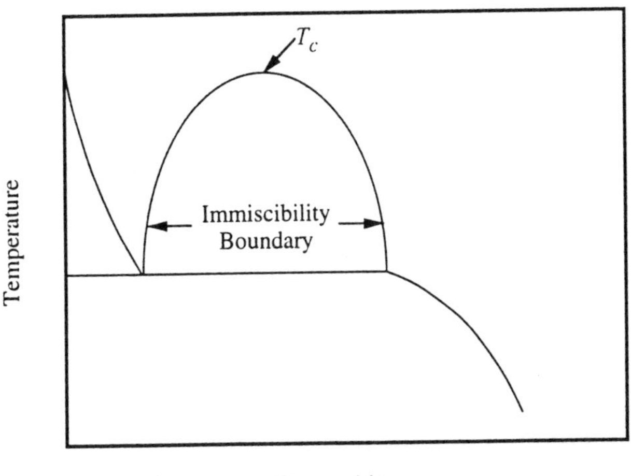

Figure 4.5 *Idealized phase diagram for a binary system exhibiting stable immiscibility*

to phase separate. If the melt is very fluid at this temperature, separation will occur rapidly and a large degree of separation will occur even if the time allowed within this temperature region is small. If the time allowed is large enough, we would expect to reach a state of complete separation into two layers. Separation into two layers would be especially likely if the melting temperature were below T_c.

The occurrence of a very fluid melt at temperatures below T_c is most likely for melts where T_c is much greater than the *liquidus*, which is the maximum temperature where crystals can exist in the melt. This condition is termed *stable immiscibility*, and is characterized by a phase diagram similar to that shown in Figure 4.5. The occurrence of phase separation in melts which exhibit stable immiscibility is very common. Examples of such systems include most of the binary systems containing an alkaline earth oxide or a transition or rare earth metal oxide and any of the most common glassforming oxides, *i.e.*, SiO_2, B_2O_3, or GeO_2. Since the fast kinetics found for stable immiscibility makes it difficult to form reproducible samples, relatively few studies have been devoted to these systems.

It is also possible for phase separation to occur in systems where T_c is below the liquidus. No immiscibility dome will be evident in the phase diagram for such systems. The liquidus curve will, however, have the typical *S-shape* shown in Figure 4.6, suggesting that a sub-liquidus region of immiscibility exist. Since the liquid is not the stable phase in this region, the occurrence of phase separation in such systems is said to be

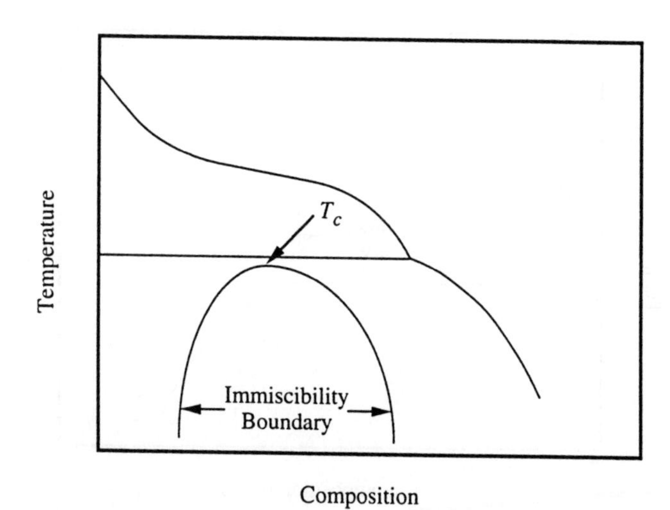

Figure 4.6 *Idealized phase diagram for a binary system exhibiting metastable immiscibility*

metastable. The viscosity of the melt at these temperatures, which are closer to the glass transformation region, will be greater and mass transport will occur at a slower rate. The progress of separation into two phases will be slowed to such an extent that the scale of the microstructure may be below that detectable by the naked eye.

Metastable immiscibility occurs in a number of very thoroughly studied systems, including the binary lithium silicate and sodium silicate systems and the ternary sodium borosilicate system. Sodium silicate and sodium borosilicate glasses are easily formed with such fine scale morphology that they appear homogeneous to the naked eye, while the slightly higher upper immiscibility temperatures of the lithium silicate melts often leads to a slightly coarser morphology and a bluish tinge due to light scattering for lithium silicate glasses. Addition of alumina suppresses the immiscibility temperature in many systems, eliminating any visible evidence of phase separation. In some cases, addition of alumina or other compounds simultaneously suppresses the immiscibility temperature while raising the glass transformation temperature, so that metastable immiscibility no longer occurs at any temperature where not prevented by slow kinetics.

Critical temperatures often follow a simple trend for systems of related compositions. The critical temperature of alkali silicate melts, for example, decreases in the order of increasing radius of the alkali ion present, such that lithium and sodium silicate melts clearly exhibit metastable immiscibility, while the existence of immiscibility in potassium silicate glasses has not been conclusively established. No evidence exists for the existence of phase separation in the rubidium or cesium silicate systems. Similar trends exist for many other systems. In general, the critical temperature within series such as the alkaline earth silicates or borates will decrease with increasing radius of the alkaline earth cation. In the former case, the suppression of the critical temperature with increasing radius of the alkaline earth cation is great enough that barium silicate melts exhibit metastable immiscibility, while the other alkaline earth silicates exhibit stable immiscibility.

DETERMINATION OF IMMISCIBILITY DIAGRAMS

Determination of immiscibility diagrams usually involves a large number of experimental measurements. The locus of the immiscibility boundary is often determined by heat treating a series of samples of constant bulk composition at different temperatures, quenching these samples, and determining if the sample is phase separated by visual observation of opalescence. The temperature of immiscibility is defined as the temperature bracketed by samples which either are or are not opalescent. The accuracy with which this temperature can be defined is determined by the temperature interval between treatments and thus depends on the number of samples used. This process must be repeated for other compositions until the complete immiscibility boundary is well defined.

A number of techniques have been used in the experimental determination of immiscibility boundaries. The method discussed above is often aided by use of *clearing studies.* If we consider the microstructure of a sample heat treated at a temperature just below the immiscibility boundary, we soon realize that the detection of phase separation may not be very easy. The lever rule predicts that only a very small quantity of the minor phase will be present, which may make detection difficult. A clearing study relies upon pretreatment of our samples at a lower temperature, where the extent of phase separation is much greater and hence more apparent. The pretreated samples are heated to the region of the immiscibility boundary, held for a specified time, and examined for the disappearance of the opalescence. The immiscibility temperature then brackets the temperatures where opalescence remains or disappears.

The detection of opalescence by the naked eye can be aided by use of a concentrated beam of light such as that produced by a laser. The intensity of the scattered light is increased over that obtained from ambient room light and the sensitivity of the determination of the

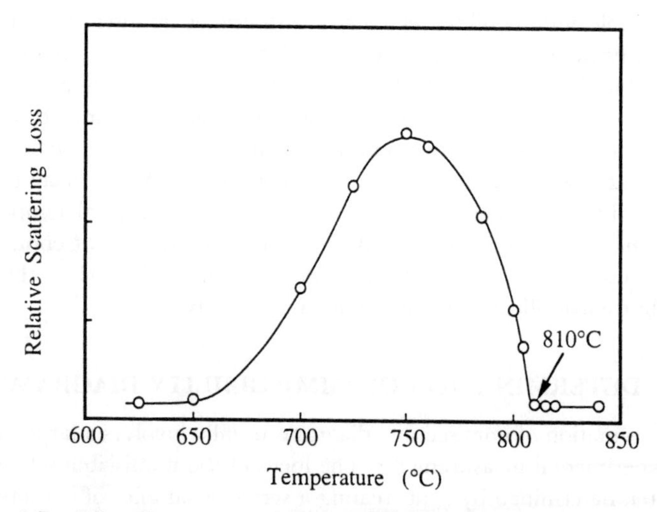

Figure 4.7 *Effect of heat treatment temperature on the scattering of light from a phase separated glass*
(Data supplied by P. B. McGinnis)

existence of light scattering is improved. The process can be made more quantitative, and a certain degree of experimental bias removed, if a detector other than the human eye is used to measure the intensity of the transmitted light, as in a spectrophotometer. Since scattering will result in loss of light intensity at the detector, a curve of scattering loss *versus* heat treatment temperature can be generated by analysis of several samples, as is shown in Figure 4.7. A decrease in scattering occurs for lower treatment temperatures due to the slow separation kinetics. The curve passes through a maximum and returns to the baseline when the sample ceases to scatter light, i.e., has been heat treated above the immiscibility temperature, which is 810 °C in this case. Glasses formed by rapid quenching from above this temperature will not scatter a significant amount of light owing to the small extent of phase separation present.

Other methods for determination of immiscibility temperatures involve direct detection of phase separation using either *X*-ray small angle scattering or electron microscopy. Both methods have been successfully used in a number of studies, but are much more instrumentation-intensive than simple visual examination. Methods based on physical property measurements, which have also been used to a limited extent, will be discussed later in this text in chapters dealing with those properties.

Determination of spinodal boundaries is much more difficult than the determination of immiscibility boundaries. Since light scattering is unaffected by the connectivity of the phases, the observation of opalescence tells us nothing about the morphology of the sample. Direct examination of the microstructure would certainly reveal the morphology and allow us to differentiate between spheres in a matrix and interconnected morphologies. Unfortunately, examination of the microstructure of glasses in regions of metastable phase separation requires the use of an electron microscope, which is both expensive and time-consuming.

Recent studies using property measurements have shown considerable promise for the detection of spinodal boundaries. A number of properties are dependent upon the degree of connectivity of one or the other of the two phases present in a phase separated glass. While these measurements will be discussed where appropriate in later chapters, an illustration using the dilatometric softening temperature will serve to establish the methodology used in these techniques. (The concept of the dilatometric softening temperature will also be discussed later.) This temperature is dependent upon the connectivity of the more viscous phase. Loss of connectivity of this phase results in a sharp decrease in the dilatometric softening temperature. The thermal expansion curves of a series of heat treated $15Na_2O$–$85SiO_2$ samples, for example, were used to establish that the spinodal boundary for this composition lies between 632 and 637 °C.

APPLICATION OF IMMISCIBILITY DIAGRAMS

Binary Immiscibility Diagrams

A simple theoretical immiscibility diagram is shown in Figure 4.2. In reality, the immiscibility region is rarely symmetrical, indicating that the regular solution model is not strictly correct for most glassforming melts. The metastable immiscibility diagram for the sodium silicate binary system, which is typical of binary glassforming systems, is shown in Figure 4.8. A calculated spinodal boundary, based on the assumption of regular solution, is indicated by the dashed line. The two measured spinodal boundary temperatures for this system lie near this line, although there is a significant difference between the calculated and experimental values for a glass containing 15 mol % soda.

The sodium silicate immiscibility diagram can be used to illustrate several applications of these diagrams. First, the locus of the immiscibility boundary separates regions of homogeneous and heterogeneous liquids.

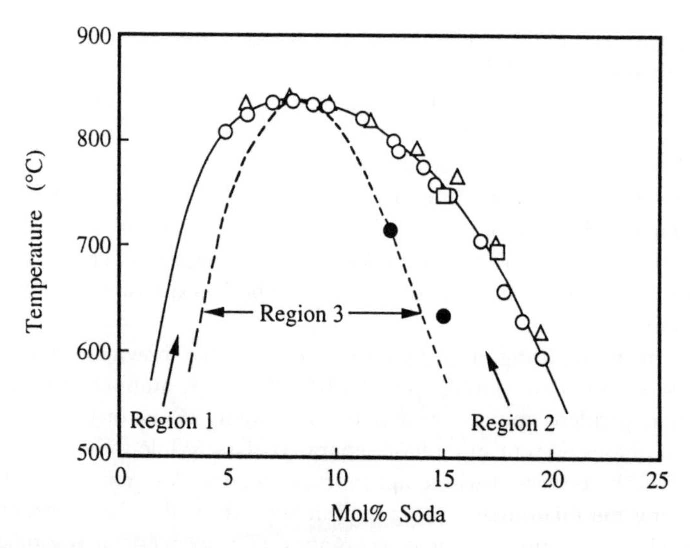

Figure 4.8 *Metastable immiscibility diagram for the sodium silicate system*

Any liquid, or melt, which is held at a temperature above this boundary will be homogeneous, and, if instantaneously quenched, would form a homogeneous glass. Any melt equilibrated at a temperature below this boundary and then quenched would form a phase separated glass. The compositions of the two vitreous phases would be given by the tie-line appropriate for the temperature used to equilibrate the melt. Any change in equilibration temperature would alter the compositions of these phases. The relative amount of each phase can be calculated using the lever rule discussed earlier.

Next, we should note that there are three regions under the immiscibility boundary. These regions represent different morphologies, or physical arrangements, of the two phases. A glass formed by instantaneous quenching from any point in the region labeled 1 will form a *droplet in a matrix*, or *droplet/matrix*, microstructure, with a matrix of the major phase, A, and droplets of the minor phase, B. Similarly, any glass formed by instantaneous quenching from any point in region 2 will also form a droplet/matrix microstructure. In this case, however, the matrix phase, which is still the major phase, will be phase B, and the droplet, or minor phase, will be phase A. Glasses instantaneously quenched from within region 3, which lies beneath the spinodal boundary, will have the interconnected structure typical of spinodal decomposition. The compositions of the two phases will still be A and B, and their relative concentrations will be given by the lever rule. The spinodal boundary only indicates the morphology of the glass and does not determine the compositions and concentrations of the phases present, which continue to be determined using the immiscibility boundary.

The reader should note that the qualifier 'instantaneously quenched' was used in discussing the results of cooling from equilibrated melts. Examination of the diagram in Figure 4.8 reveals that a decrease in temperature often involves passing from one region to another during the cooling process. All melts, for example, formed for compositions lying between the outermost limits of the immiscibility boundary, must pass through the miscibility gap during cooling from temperatures above the immiscibility boundary. If sufficient time is allowed at temperatures within the miscibility gap, phase separation will occur during cooling, and an originally homogeneous melt will still yield a phase separated glass. The extent of phase separation, *i.e.* the size of the separated regions and their compositions, will be determined by the interaction of the cooling rate and the kinetics of phase separation for that particular composition.

The lack of observation of visible light scattering or opalescence is sometimes taken as an indication that a melt was cooled fast enough to prevent any separation. This is rarely true unless extremely fast quenching methods are used. Bulk glasses formed from melts which lie above a metastable immiscibility region are virtually always phase separated on a microscopic scale. Unless confirmed by either electron microscopy or *X*-ray small angle scattering, the statement that such a glass is homogeneous should be treated with considerable skepticism.

Further examination of this diagram reveals that most melts will pass through two of the three regions under the immiscibility boundary during cooling. The only melt which can be cooled directly into the spinodal region (region 3) without first passing through either of the nucleation and growth regions (regions 1 or 2) has the composition of the *critical composition, i.e.*, the composition of the melt with an immiscibility temperature equal to the critical temperature. Other melts with compositions bounded by the spinodal boundary will begin to separate by a nucleation and growth mechanism as they are first cooled below the immiscibility boundary and then convert to a spinodal mechanism as they are cooled further to temperatures below the spinodal boundary. The result can be a complex morphology, with elements of both droplet/matrix and interconnected structures.

If the melt composition lies toward the edges of the immiscibility region, it may be possible to avoid passing through the spinodal region, as, for example, would be the case for sodium silicate melts containing either 3 or 18 mol % soda. Glasses formed from these melts should have a droplet/matrix morphology. However, there are mechanisms which can eventually lead to redistribution of the phases as the spheres grow, which is called *coarsening*, and yield interconnected minor phases under certain conditions.

Reheating of glasses to temperatures above T_g and within the immiscibility region can often alter their microstructures. If we consider a sodium silicate glass containing 15 mol % soda formed by rapid cooling, we might find that the microstructure has been effectively frozen in at temperatures inside the immiscibility boundary, but outside the spinodal boundary. Reheating the glass to a temperature just below the spinodal boundary, and holding at this temperature for an extended time, will allow conversion of the microstructure from the initial droplet/matrix structure to an interconnected structure, with changes in the compositions and concentrations of the two vitreous phases as demanded by the new tie-line. Subsequent heating and equilibration at a temperature above the spinodal boundary would cause a reversion to a droplet/matrix microstructure, with *necking* of narrow regions in the interconnected structure to form spheres.

Only a few other binary immiscibility diagrams are well established. With the exception of the lithium and barium silicate and lead borate systems, all other glassforming binary immiscibility systems are either within the realm of stable immiscibility or are not well established. The most controversial systems include all five alkali borate systems, where considerable argument over the presence of phase separation exists. At this time, it appears that earlier reports of immiscibility regions in these systems are incorrect with the possible exception of the lithium borate system, where some questions regarding the presence of immiscibility still exist. In many other cases, *e.g.*, lead silicates and germanates, phase separation is known to exist, but the immiscibility boundary has not been established. Metastable immiscibility has been reported for the alumina and gallia silicate systems, but is not well established. The existence of immiscibility in the B_2O_3–SiO_2 binary system is also in question.

Ternary Immiscibility Diagrams

Although only a few binary glassforming systems definitely exhibit metastable immiscibility, a large number of ternary systems clearly contain regions of immiscibility. These systems can be separated into two major categories: systems containing only one glassformer and those containing two glassformers. The former category includes the commercial soda–lime–silicate system, which serves as the basis for the great bulk of common consumer products, while the latter category contains

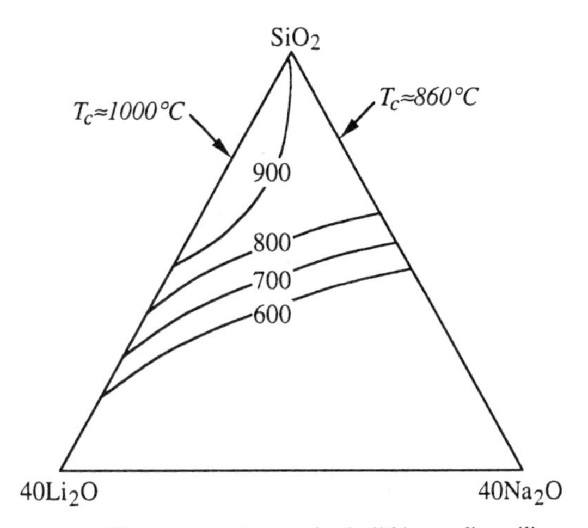

Figure 4.9 *Immiscibility contour diagram for the lithium sodium silicate system*

the sodium borosilicate system, which serves as the basis for a large number of technical glass products. Many of these systems consist of silica combined with two modifier oxides, of which at least one displays metastable immiscibility with silica, while the other may either display stable immiscibility with silica or be a component in a phase separation-free system with silica.

Depiction of immiscibility regions in ternary systems is somewhat more confusing than in binary system. Since the composition is normally displayed using triangular coordinates, temperature must be displayed on a fourth axis, which is perpendicular to the plane of the composition diagram. Since the immiscibility region is actually a dome in a ternary system, isothermal contours from that dome are projected onto the plane of the composition diagram. These contours represent the position of the immiscibility boundary at the indicated temperatures. An example of such a diagram for the lithium–sodium silicate system is shown in Figure 4.9.

The spinodal boundary is rarely shown on ternary immiscibility diagrams. Since this boundary is also represented by a dome, similar contour lines could be drawn. However, since the spinodal boundary is rarely known for these systems, it is usually neglected in the presentation of ternary immiscibility regions.

In many cases, a more pragmatic version of the diagram shown in Figure 4.9 is used, as is shown in Figure 4.10 for the same system. The only contour presented for the immiscibility boundary is that determined at a low temperature, often near T_g, which represents the *practical limit of*

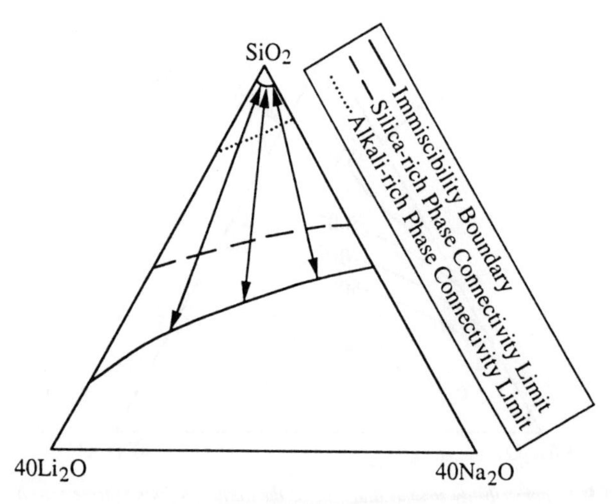

Figure 4.10 *Tie-line diagram for the lithium sodium silicate system*

immiscibility. All melts which lie within the region enclosed by this contour will lie within the immiscibility region. The lack of other contour lines now allows inclusion of a contour for the spinodal boundary at the same temperature, as well as representative tie-lines. Since knowledge of tie-lines is essential to the determination of the compositions and concentrations of the phases present, these diagrams also serve a very useful, albeit different from that of the full immiscibility dome diagram, purpose.

The tie-lines shown in Figure 4.10 display a fan-shaped pattern, with the base of the fan in the silica corner of the composition diagram. This pattern is typical of ternary systems where both modifier oxides exhibit immiscibility with the network oxide. These systems include the Li_2O–RO, Na_2O–RO, and BaO–RO silicate systems, where RO is any alkaline earth oxide. A different pattern occurs for systems where immiscibility occurs for only one of the modifier oxides in its binary system with the network oxide. In this case, the immiscibility region is restricted to the area near the binary where immiscibility does occur and the tie-lines do not extend to the network oxide corner of the diagram.

Immiscibility diagrams for ternary systems containing two network oxides are more complex in appearance than those discussed above, but are based on exactly the same principles (see Figure 4.11). Contours are drawn to indicate the position of the immiscibility boundary at the indicated temperatures. Diagrams containing only the practical limits of immiscibility are also frequently published. The position of the critical

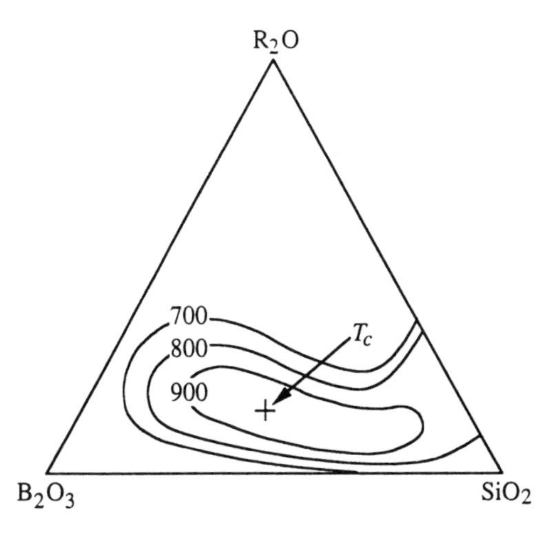

Figure 4.11 *Immiscibility contour diagram for an arbitrary alkali borosilicate system*

composition and the critical temperature, *i.e.*, the position of the maximum of the dome, are often indicated on these diagrams. A thorough determination of a series of boundary contours is very time-consuming and hence has been carried out for only a few systems.

Determination of tie-lines in these systems is more difficult, since they do not necessarily follow as simple a pattern as those, for example, in the lithium–sodium silicate system. Tie-lines can also rotate with temperature, so that the lines at 1000 °C are not necessarily co-linear with those at 800 °C. Since chemical analysis of phases which are so small in size is very difficult, techniques based on property measurements have been developed to define tie-lines. Measurement of T_g, which will be constant for all samples lying on the same tie-line, have proven particularly useful for determining both tie-line directions and end member compositions.

The extent of the immiscibility region for alkali borosilicate systems decreases with increasing atomic number of the alkali present. Lithium borosilicate melts border on stable immiscibility, so that glasses produced from ternary compositions are quite opalescent and, occasionally, are even opaque due to the large sizes of the separated regions. Immiscibility in sodium borosilicate melts lies entirely within the metastable region, so that the glasses formed from these melts usually appear homogeneous to the naked eye. Since only a very small region of phase separation is found in the potassium borosilicate system, most glasses formed in this ternary compositional system are homogeneous, as are the rubidium and cesium borosilicate glasses. Virtually all other ternary borosilicate systems exhibit stable immiscibility, with very large immiscibility regions covering most of the region of glass formation in these systems. Small additions of alumina as a fourth component generally suppress the immiscibility temperatures in alkali borosilicate systems.

Phase separated alkali borosilicate glasses tend to consist of a silica-rich phase and an alkali borate rich phase. Several studies have shown that impurities, including transition metals and hydroxyl, are concentrated in the alkali borate rich phase. Since the alkali borate rich phase is much more readily dissolved in HCl than is the silica-rich phase, glasses with interconnected microstructures can be leached to remove the alkali borate rich phase and leave a silica skeleton. Subsequent heat treatment can cause the collapse of the porous silica skeleton to form a fully dense (or very nearly so) glass with a very high silica content and a low impurity content. These porous materials can also be used as filters and are currently the subject of considerable interest as possible gas separation membranes.

SUMMARY

Liquid–liquid immiscibility is a common phenomenon in glassforming melts. Understanding of immiscibility is based on the thermodynamics of regular solutions. The separation process can yield glasses with either droplet/matrix or interconnected microstructures. Immiscibility diagrams are used to indicate the regions of immiscibility and the compositions and concentrations of the co-existing phases. A number of commercial products are based on phase separated glasses, especially those in the sodium borosilicate system.

Chapter 5

Structures of Glasses

INTRODUCTION

Upon first consideration, the term 'structures of glasses' appears to be an oxymoron. How can materials defined by a lack of long range, periodic structure have structures which are characteristic of specific compositions? On the other hand, we know that the properties of three different samples of glasses of the same nominal composition, produced independently in three different laboratories, and annealed in the same manner, will be identical within reasonable limits. Our basic understanding of the solid state then indicates that these glasses have, if not identical, at least very similar structures. It follows that a lack of long range, periodic structure does not, in and of itself, preclude the existence of structure at a level that will control the properties of material.

Early discussions of glass structures centered on silicate glasses, especially vitreous silica and alkali silicate glasses. The first models for glass structures were based on the structures of silicate crystals. The *microcrystal* approach suggested that glasses are simply masses of very small, or micro, crystals. The small size of these crystals can be used to explain the lack of structure in the *X*-ray diffraction pattern. Lebedev and others in Russia favored a somewhat different version of the microcrystalline model, which they termed the *crystallite* model. Their crystallites differ from microcrystals in that the structures are deformed versions, *i.e.*, not perfect lattices, of those of the crystals and are not merely smaller sized versions of normal crystals. A glass then was assumed to consist of crystallites connected by amorphous regions (similar to grain boundaries). The average composition of the glass is obtained by use of the appropriate concentrations of two or more crystalline phases, whose composition can be determined from the phase diagram for that system.

Table 5.1 *Zachariasen's rules for glass formation in simple oxides*

(1) Each oxygen atom is linked to no more than two cations
(2) The oxygen coordination number of the network cation is small
(3) Oxygen polyhedra share only corners and not edges or faces
(4) At least three corners of each oxygen polyhedra must be shared in order to form a three-dimensional network

Modified rules for complex glasses

(5) The sample must contain a high percentage of network cations which are surrounded by oxygen tetrahedra or triangles
(6) The tetrahedra or triangles share only corners with each other
(7) Some oxygens are linked only to two network cations and do not form further bonds with any other cations

The crystallite theory suggests that the properties of glasses in a specific system should be associated with the phase diagram for that system. Many studies have been interpreted using arguments based on detection of inflections in property/composition curves at compositions identical to those of specific crystalline phases. In general, these interpretations are questionable, with little evidence to support the existence of residual crystalline behavior in most glasses. Other theories which do not invoke the existence of crystallites also occasionally predict changes in glass property trends at compositions which correspond to specific crystalline phases. From the perspective of the glass scientist, however, it seems that the formation of a new crystalline phase is forced by the same changes in network connectivity which lead to changes in glass structures and not that the existence of the crystalline structure forces changes in glass structures.

The most commonly used models for glass structures today are based on the original ideas of Zachariasen, and are grouped under the term *random network theory*. Zachariasen's classic paper, which never used the term random network, was not intended as a discussion of structural models, but as an explanation for glass formation tendencies. *Zachariasen's rules* for glass formation, however, have been expanded through wide usage into a set of rules for formulating models for glass structures. These rules, which are summarized along with three modified rules for more complex systems in Table 5.1, simply state the conditions for the formation of a continuous three-dimensional network, with no indication of the degree of long range order of that network. In fact, these rules adequately describe the structures of many crystalline phases, particularly those in silicate systems!

The structural model offered by Zachariasen provides an approach for describing network structures, whether or not they are glasses. In order to explain glass formation, Zachariasen added the requirement that these networks be distorted in some manner such that long range periodicity is destroyed in order for a glass to form. These distortions can be achieved by variations in bond lengths or bond angles, and by rotation of structural units about their axes.

Most of the structural models found in the literature today actually only address the formation of a network. Very few directly address the issues of bond angle distributions, bond rotations, and bond length variations inherent to a random network model. Many of these models only attempt to explain trends in one property or type of spectra, ignoring masses of other data for the same system which often cannot be explained using the proposed model. As a result, one should be very cautious in accepting any structural model for glasses without a thorough understanding of the limitations of the model and, unfortunately, possible bias on the part of those proposing the model.

FUNDAMENTAL LAW OF STRUCTURAL MODELS

Historically, models for 'structures' of glasses have been based on a desire to explain trends in properties. The original 'random network model' attributed to Zachariasen was based solely on property data, *i.e.*, glass formation behavior. While the work of Warren and co-workers which soon followed the proposal of the concept of a random network supported this model, Warren's work was not the source of the model. In fact, virtually all of the basic models for the structures of oxide glasses were originally proposed as means of justifying the trends observed in the properties of simple compositional series of glasses. Although modern spectroscopic methods (Raman and infrared spectroscopy, nuclear magnetic resonance, neutron diffraction, electron spin resonance, EXAFS, XPS, *etc.*) now allow researchers to propose models for new glasses solely on the basis of spectroscopic data, these models must still pass the time-honored test of explaining all of the trends observed in the properties of those glasses before such models can be accepted as truly representing the structure of the material. It should also be recognized that these spectroscopic methods are best suited for examining the local structure about an ion, with some extension to the intermediate range of order, and, in general, are quite incapable of application to materials where the properties are controlled by microstructure rather than short to intermediate range atomistic order.

As a further guide to the testing of any proposed model, the author now proposes the *Fundamental Law of Structural Models*:

Table 5.2 *Required elements for any complete structural model for glasses*

(1)	Coordination number of all network cations
(2)	Distribution of bond angles and rotations
(3)	Connectivity of all network units
(4)	Dimensionality of the network
(5)	Nature of any intermediate range order
(6)	Morphology
(7)	Field strength, bond strength, site specific bonding
(8)	Nature of the interstitial or free volume
(9)	Role of minor constituents, impurities, defects, *etc.*

> 'No model can be considered to be valid unless that model can explain **ALL** of the available data.'

Any model failing to consider and offer an explanation for all of the existing data (not just that obtained in one particular study) must be considered suspect and incomplete. It follows that an inverse statement of the fundamental law of structural models can be formulated:

> 'Any structural model which can only explain some of the available data and fails to explain the rest is inherently flawed.'

ELEMENTS OF STRUCTURAL MODELS FOR GLASSES

A complete structural model for any glass must contain a number of elements, as indicated in Table 5.2. A number of these elements will be discussed in some detail before proceeding to a discussion of structural models of a number of common glasses.

Coordination of the Network Cations

The coordination numbers of all cations which occupy sites in the vitreous network, *i.e.*, the *building blocks* which constitute the network, provide the most basic element of any model for glass structures. Since these building blocks are usually well defined structures such as tetrahedra or triangles, they exhibit order at the level of several associated atoms or ions. Since this order only extends over a very short distance, it is termed *short range order*.

It follows that the starting point for any model for the structure of a vitreous materials must lie in a determination of the coordination unit for the high field strength cations, which serve as the network-forming cations in the material. Traditionally, we estimate the possible coordination number of each cation using the radius ratio or by consideration of the coordination states usually observed for these cations in other materials. On this basis, we can state that silicon is almost always tetrahedrally coordinated in silicate glasses, that boron can potentially exist in either three- or four-fold coordination, and that aluminum and germanium might reasonably be expected to exist in either four- or six-fold coordination in oxide glasses. Potential coordination states for virtually any ion can be established by similar considerations.

Fortunately, we no longer have to rely solely on geometric arguments and analogies based on crystalline materials as sources for models for the coordination state of the network cations in glasses. Many modern spectroscopic methods can directly determine these states. The advent of magic angle spinning nuclear magnetic resonance (MAS-NMR) has virtually removed the element of speculation from models for the coordination state of Si^{4+}, Al^{3+}, B^{3+}, P^{5+}, and other ions in glasses. While other methods can also directly determine coordination states, MAS-NMR has proven to be especially useful, as illustrated by the case of aluminum, where it has been established that aluminum ions not only exist in the traditional four- and six-fold coordination states, but also often exist in five-fold coordination in glasses.

Bond Angle Distributions

After determination of the building blocks, we next must specify how these building blocks are connected, including the distributions in bond angles and rotations which introduce randomness into the structure. Unfortunately, experimental determination of these distributions is very difficult and has been carried out for very few glasses. As a result, while the concept of bond angle distributions remains an important part of glass structural models, very little quantitative information is available.

Network Connectivity

Descriptions of glass structures also always address the number and arrangement of *bridging* and *non-bridging* bonds which link each of the building blocks to their neighbors, *i.e.*, the *connectivity* of the network. Most models for vitreous networks only consider connectivity as evidenced in the concentration and distribution of *non-bridging oxygens*, or *NBO*, *i.e.*, those oxygens which do not link network polyhedra. Spectral techniques such as MAS-NMR, Raman spectroscopy, and XPS have proven to be invaluable in establishing the distribution of non-bridging oxygens on the silicon–oxygen and aluminum–oxygen tetrahedra of glasses.

Network connectivity, however, should not be considered solely in terms of NBO concentrations. The coordination of the network cations is also an important factor in determining the connectivity of the network. As an example, consider the well-known case of alkali borate glasses (discussed in detail later in this chapter). It is generally believed that the addition of small concentrations of alkali oxide to these glasses converts boron ions from three- to four-fold coordination, without the formation of NBO. Network polyhedra are thus converted from triangles to tetrahedra, with an accompanying increase in the number of bonds connecting each of these converted polyhedra into the network.

Unfortunately, while it is easy to recognize the importance of network connectivity in discussing any structural model, it is more difficult to define this term in a quantitative manner. The author proposes that the average number of bridging bonds per network unit be termed the *connectivity number* and used as a measure of the connectivity of any glass. The use of the connectivity number will be illustrated throughout the following discussion.

Dimensionality

Since networks can exist in both two and three dimensions, we must also specify the *dimensionality* of the network, which dramatically affects the ease with which the structure can be spatially distorted. Dimensionality of a network is related to, but certainly different from, connectivity. The connectivity number only considers the average number of bridging anions per network unit and does not consider how they are arranged. Vitreous boric oxide, for example, consists of a network of boron–oxygen triangles, while vitreous silica consists of a network of silicon–oxygen tetrahedra. Vitreous boric oxide can be considered to have a planar structure, existing in three dimensions in much the same way a drawing on a piece of paper becomes three-dimensional when the paper is crumpled into a ball. Vitreous silica, on the other hand, is a true three-dimensional network. This difference in dimensionality can be used to explain the radical difference in the glass transformation temperatures (T_g of B_2O_3 is $\approx 260\,°C$, while that of vitreous silica is $\approx 1100\,°C$) of these two compositionally simple glasses. Other glasses may even exist as a tangled structure of long-chain polymers, which basically have a dimensionality of one, and again occupy three dimensions in the same way as does a ball of yarn. Dimensionality is especially important when polymeric glasses are drawn into fibers, where the structure can become oriented rather than random.

Dimensionality is also difficult to express in a quantitative manner.

Obviously, the connectivity number does not properly describe the dimensionality. The issue is further complicated by the need to include some representation of possible orientation effects, especially for polymeric glasses and possibly for layer structure glasses. Since no quantitative method for describing dimensionality exists, the development of such a method should be considered as a challenge for the community of glass scientists to create a proper method for describing these structures.

Intermediate Range Order

The concept of dimensionality leads to the possibility of somewhat longer range order than that found within the basic building block. These blocks may be connected into slightly larger units which have a more ordered arrangement than that predicted by a purely random connection scheme. We may find rings or chains of building blocks connected in a manner closely resembling those found in crystals, but yet not extending over significant distances within the structure. Such units are said to provide a degree of *intermediate range order* to the structure.

Although intermediate range order is well established in certain systems, its existence in others is more speculative. A trend has arisen in recent years to find evidence for intermediate range order in every spectrum and bit of datum presented. In many cases (probably a majority), the evidence for intermediate range order is beyond the capability of the technique used. The ability to provide an approximate fit to a spectral curve by assuming the existence of intermediate range units does not prove that these units actually exist. Even if the fit is in excellent agreement with experimental data, it must be remembered that no intermediate range unit can be demonstrated to exist until it has been shown that no other model will yield an equal or better fit. The problem of proving the uniqueness of intermediate range order models has not been solved and does not appear to be likely to be solved in the near future. Until it becomes possible to prove that a proposed model provides a unique structure and not simply a possible structure, all models for intermediate range order must be treated only as potential descriptions of a network and not as absolutely established.

Morphology

Morphology resulting from phase separation is frequently neglected in proposing glass structural models. It is not surprising that the effect of morphology is routinely neglected in spectroscopic studies, where the data produced are not sensitive to morphology, *i.e.*, the spectra obtained are basically just the sums of the spectra of the phases present. It is, however, very surprising that morphology is so often ignored in studies of properties which are highly dependent upon the compositions and connectivity of the phases present, *e.g.*, the glass transformation temperature, transformation range viscosity, electrical conductivity, chemical durability, *etc.* A number of frequently quoted structural models for glasses in systems exhibiting a large immiscibility region are incorrect.

Properties of Specific Ions

Other elements of glass structures must be included in order to discuss more subtle variations in properties and various spectra. The *field strength* of both network-forming and network-modifying cations must be included in discussions of trends in properties with glass compositional variations. Since many modern studies include variations in the identity of the anions which link the structure, their field strengths and ionic radii are also important. The atomic radii of mobile cations, anions, and atoms and molecules must be considered in structural models seeking to explain transport properties.

Interstitial/Free Volume

The arrangement of the coordination units in space to form *interstices* is virtually always ignored in structural models. Knowledge of the *interstitial volume*, or *free volume*, in glasses is vital to an understanding of any diffusion-based process, as well as detailed explanations for the behavior of volume-dependent properties such as the density, refractive index, and thermal expansion coefficient. Unfortunately, it is very difficult to study the interstitial/free volume. How does one study what is not there? Spectroscopic studies are beginning to address this question *via* consideration of possible ring structures in the network. Of course, the ring structures only form the faces of the interstices and do not, therefore, directly address the question at hand.

Perhaps the best approach to this problem lies in the study of inert gas solubility and diffusion in glasses. Inert gas atoms can be used as probes of the interstitial regions of glasses. It is also possible that computer simulations of glass structures will aid in understanding the 'empty space' which we all know must exist in these network structures. In any case, it is certain that a full understanding of the structure of a glass has not been obtained until we can confidently predict such basic properties as the density.

Minor Constituents

Finally, the role of minor constituents, impurities, and defects must be included to characterize a vitreous structure completely. In many cases, the local structure around these sites plays an important role in determining certain properties of glasses. If so, this site structure should be included in the complete structural model for the glass.

In particular, we find that most studies of glass structures ignore the possible effects of minor components and impurities. This problem is particularly evident in studies of unusual glassforming systems, where unsuspected reactions with crucibles may occur. The very existence of glass formation in a given system may depend upon dissolution of some of the crucible material into the melt, especially when melts are prepared in oxide crucibles. In general, glasses prepared in oxide crucibles, especially porcelain or silica, should be considered suspect unless evidence is provided that the crucible is not a factor in glass formation. Since the structural models proposed for these systems assume that no significant impurities exist in the glass, they can be completely misleading.

The effect of impurities is especially important when the impurity is water, which is usually present in the form of hydroxyl. Since hydroxyl has a very large effect on the properties of oxide glasses, it must be included in a complete structural model. While usually ignored in such models, variations in the hydroxyl contents of a series of specimens may often be the controlling factor in determining their properties, especially if the property would vary only slightly for glasses of identical hydroxyl contents. Since the hydroxyl content is usually uncontrolled and can vary from melt to melt, most experimenters choose to ignore it completely. At the very least, the hydroxyl content of the samples should be discussed in the paper so that the reader can consider it in forming an opinion of the validity of the study.

Comments Regarding Glass Structure Models

Unlike crystalline materials, no single structure exists for any given glass composition, or even spatially within a given sample. Instead, we must always discuss an idealized structure, which is known to vary within certain bounds throughout the specimen. These structural models will prove to be much more useful if they are always based on a desire to explain all of the observations, including those of others, from both spectral and property studies, rather than restricted to the data of the authors of the paper. Anyone purporting to offer a model for a glass structure must consider all of the evidence and not just their own findings. Until this approach becomes common practice, the literature will continue to be filled with conflicting models and a true understanding of the structures of glasses will be impossible.

STRUCTURAL MODELS FOR SILICATE GLASSES

Discussions of the structural models for oxide glasses almost always begin with vitreous silica and the alkali silicate glasses. Structural models for most other silica-based glasses are derived from those for these systems, as is most of the terminology used in discussing glass structural models in general.

Vitreous Silica

The structure of *vitreous silica* is readily described by the network structural rules of Zachariasen (Table 5.1). The silicon–oxygen tetrahedron, with a coordination number of 4, serves as the basic building block for the network, as required by the second of Zachariasen's rules. Since these tetrahedra have a high degree of internal order, the short range order of the glass is preserved. These tetrahedra are linked at all four corners (rules 3 and 4) to form a continuous, three-dimensional network. Each oxygen atom is shared between two silicon atoms, which occupy the centers of linked tetrahedra. Disorder in this structure is obtained by allowing variability in the Si–O–Si angle connecting adjacent tetrahedra. Additional disorder is introduced by allowing rotation of adjacent tetrahedra around the point occupied by the oxygen atom linking the tetrahedra, and by rotation of the tetrahedra around the line connecting the linking oxygen with one of the silicon atoms. Since the Si–O–Si angle and the rotations are described by distributions of values rather than the single values found in crystal lattices, no long range periodicity exists.

A two-dimensional representation of such a structure is shown in Figure 5.1, where the fourth oxygen, which would sit directly above the small silicon ion, is not shown. Note the existence of rings consisting of three or more tetrahedra and interstices of various sizes and shapes.

Diffraction studies indicate that the shortest Si–O distance in this structure is $\approx$0.162 nm and that the shortest O–O distance is $\approx$0.265 nm. These distances are consistent with those found within silicon–oxygen tetrahedra in crystalline forms of silica and in silicate minerals. These distances exhibit very small variations, illustrating the high degree of order within the short range represented by the basic tetrahedral

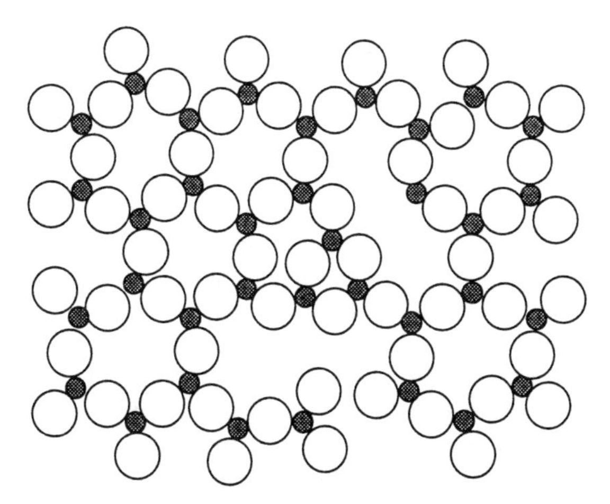

Figure 5.1 *Schematic drawing of a two-dimensional structure for a pure glassformer. A fourth oxygen would be located above each cation in vitreous silica*

building block. The next distance in the structure, which should be between silicon atoms in the centers of linked tetrahedra, displays a considerable range of values, however, clustered around a distance of 0.312 nm as a result of the distribution in Si–O–Si angles. Even broader distributions are found for atom pair distances such as the silicon–second oxygen ($\approx$0.415 nm) and oxygen–second oxygen ($\approx$0.51 nm) distances.

These distributions of atom–atom distances can be made to fit by assuming a distribution in the Si–O–Si bond angles. The maximum in this distribution occurs at $\approx 144°$, with a range in angles from 120 to 180°. The distribution is relatively narrow, with most of the angles lying within $\pm 10\%$ of 144°. While the exact distribution of Si–O–Si angles may be subject to some disagreement due to the limitations of the experimental data, this general description appears adequate for a basic understanding of the structure of vitreous silica.

The structure of vitreous silica has regions of highly stressed bonds and defects such as oxygen vacancies, represented by Si–Si bonds, and peroxy defects, represented by Si–O–O–Si bonds. Additional defects occur at impurity sites, especially those associated with bound hydrogen species such as SiOH and SiH.

Alkali Silicate Glasses

Alkali silicate glasses containing large concentrations of alkali oxides can be easily produced by melting silica with alkali carbonates or nitrates. Glasses containing less than ≈ 10 mol % alkali oxide are considerably more difficult to melt due to their high viscosities. Metastable immiscibility occurs in the lithium and sodium silicate systems, with immiscibility limits extending to ≈33 mol % for the lithium silicate system and to ≈20 mol % for the sodium silicate system. There is little evidence for liquid–liquid immiscibility in the other alkali silicate systems (K, Rb, Cs), although there is some controversy concerning a report of a small region of metastable immiscibility in the potassium silicate system.

If we avoid the regions of metastable immiscibility, we find that the addition of any alkali oxide to silica to form a binary glass results in major reductions in the viscosity of the melt (many orders of magnitude) and the glass transformation temperature (≈500 K). The densities and refractive indices and the thermal expansion coefficients of the glasses increase with increasing alkali oxide concentration and with the atomic number/mass of the particular alkali present. The electrical conductivity of the alkali silicate glasses, which is due to diffusion of alkali ions, increases by orders of magnitude with increasing alkali oxide content.

These trends in properties are due to the formation of non-bridging oxygens, which decrease the connectivity of the melt. The structure can be viewed as a network of silicon–oxygen tetrahedra with occasional breaks in connectivity due to the non-bridging oxygens. Each non-bridging oxygen must be associated with a nearby alkali ion to maintain local charge neutrality. These alkali ions occupy the interstices in the network, reducing the unoccupied free volume of the structure. The concentration of non-bridging oxygens increases, and the concentration of bridging oxygens decreases, directly in proportion to the alkali oxide content, until a network can no longer be maintained. A two-dimensional drawing of such a structure is shown in Figure 5.2, which includes alkaline earth ions as well as alkali ions.

A more quantitative model for these glasses can be obtained if we calculate the concentrations of bridging and non-bridging oxygens per silicon–oxygen tetrahedron as a function of the alkali oxide concentration. A common nomenclature system known as the *Q_n notation* expresses the concentration of bridging oxygens per tetrahedron by varying the value of the subscript *n*. A tetrahedron fully linked into the network *via* four bridging oxygens is designated as a Q_4 unit, while an isolated tetrahedron with no bridging oxygens is designated as a Q_0 unit. The value of *n* is thus equal to the number of bridging oxygens on a given tetrahedron. Determination of the concentrations of each of the five possible Q_n units therefore characterizes the connectivity of the structure.

Calculation of Q_n concentrations involves assumptions based on our preconceived notion of the structure of a particular series of glasses. In

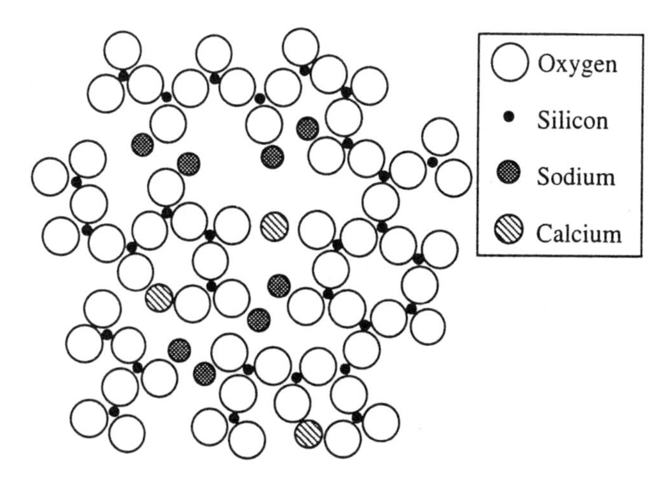

Figure 5.2 *Schematic drawing of a two-dimensional structure for a soda–lime–silicate glass. A fourth oxygen would be located above each silicon in the three-dimensional structure*

this case, we begin with vitreous silica, which is defined as consisting of 100% Q_4 units. We now assume that addition of a generic alkali oxide, expressed as R_2O, where R is the cation, creates non-bridging oxygens. We further assume that only Q_3 units will be produced until every tetrahedron in the network is linked by three bridging oxygens and one non-bridging oxygen, *i.e.*, Q_3 = 100%. Further additions of R_2O are assumed to produce Q_2 units, with a concurrent decrease in the number of Q_3 units, until all units are Q_2 tetrahedra. This series of assumptions is continued until only Q_0 units exist.

This simple Q_n model further assumes that only two types of Q_n units will exist in any given glass. All of the oxygens in the structure will be divided between the two types of units present in that glass. Using these assumptions, we can now write an expression equating the total number of oxygens present in the glass formulation with the oxygens distributed between the Q_n units. If we write the general glass compositional formula

$$xR_2O \cdot (100 - x)SiO_2 \tag{5.1}$$

where x is the mol % of R_2O in the glass, we find that the total oxygen content is given by the sum of the mol % of each component multiplied by the number of oxygens in that component. For our generic alkali silicate glass, therefore, since there is one oxygen in each of x mol % of R_2O and two oxygens in each mol % of SiO_2

$$\text{number of oxygens} = x(1) + (100 - x)(2) \tag{5.2}$$

We now must distribute the oxygens among the Q_n species in the glass. Each tetrahedron in a Q_4 unit shares four oxygens with its neighbors. Since each bridging oxygen can be considered as a part of each of two tetrahedra, we can only count half of each oxygen as a part of our Q_4 unit, which therefore has 2.0 oxygens per tetrahedron. A Q_3 unit contains three shared, bridging oxygens and one unshared, non-bridging oxygen, which, since unshared, is entirely a part of our unit, and therefore contains 2.5 oxygens per tetrahedron. It follows that a Q_2 unit contains 3.0 oxygens per tetrahedron, a Q_1 unit contains 3.5 oxygens per tetrahedron, and a Q_0 unit contains 4.0 oxygens per tetrahedron. Since all of the oxygens must be partitioned among these units, we can write an expression for the total number of oxygens present as the sum of the product of the number of each type of unit multiplied by the number of oxygens per tetrahedron for that unit. In the present case, since we have assumed that only two types of units coexist in the glass, and that we begin with vitreous silica, where we have only Q_4 units, which are replaced by Q_3 units as the value of x increases, we can write an expression for the total number of tetrahedra, $Q_4 + Q_3$, which must equal the number of tetrahedra expressed by the glass formula, or $100 - x$, *i.e.*,

$$Q_4 + Q_3 = 100 - x \tag{5.3}$$

If we now replace Q_4 by the quantity $100 - x - Q_3$, from Equation 5.3, we find that

$$\text{number of oxygens} = [(100 - x) - Q_3](2.0) + Q_3(2.5) \tag{5.4}$$

Finally, setting Equation 5.2 equal to Equation 5.4, we obtain the expression

$$x(1) + (100 - x)(2) = [(100 - x) - Q_3](2.0) + Q_3(2.5) \tag{5.5}$$

which can be solved to yield the number of Q_3 units expressed by the generic composition in terms of x, the mol % of alkali oxide. The result

$$Q_3 = 2x \tag{5.6}$$

can be combined with Equation 5.3 to yield an expression for the number of Q_4 units, or

$$Q_4 = 100 - 3x \tag{5.7}$$

Once we know how Q_4 and Q_3 vary with the concentration of alkali oxide, we can calculate several other useful numbers. First, we can determine the concentration of alkali oxide needed to eliminate completely all fully bridged Q_4 units from the structure. Setting $Q_4 = 0$ in

Equation 5.7, we find that, if x = 33.3 mol %, our glass contains only Q_3 units. Furthermore, the average number of oxygens per tetrahedron is given by the total number of oxygens in each type of Q_n unit divided by the total number of tetrahedra in the generic glass composition, or

$$\frac{\text{oxygens}}{\text{tetrahedron}} = \frac{(100 - 3x)(2.0) + 2x(2.5)}{(100 - x)} \tag{5.8}$$

If we wish to know the fraction of oxygens which are non-bridging, we simply divide the number of Q_3 units ($2x$), which each contain one NBO, by the total number of oxygens, *i.e.*, the number of NBO plus the number of bridging oxygens, BO, as given by Equation 5.2, or

$$\frac{\text{NBO}}{\text{NBO} + \text{BO}} = \frac{2x}{x(1) + (100 - x)(2)} = \frac{2x}{200 - x} \tag{5.9}$$

We might also wish to know the average number of NBO per tetrahedron, which, since there are no NBO in a Q_4 tetrahedron and one NBO in a Q_3 tetrahedron, is given by

$$\frac{\text{NBO}}{\text{tetrahedron}} = \frac{2x(1)}{(100 - x)} = \frac{2x}{100 - x} \tag{5.10}$$

The connectivity of the structure might be discussed in terms of the average number of bridging corners per tetrahedron, which would be given by the contributions from the Q_4 (four bridging corners) units plus those from the Q_3 (three bridging corners) units divided by the number of tetrahedra, or

$$\frac{\text{bridges}}{\text{tetrahedron}} = \frac{(100 - 3x)(4) + 2x(3)}{(100 - x)} = \frac{400 - 6x}{100 - x} \tag{5.11}$$

In this particular case, the number of bridges per tetrahedron is equal to the connectivity number discussed earlier. Since the connectivity number will decrease monotonically with increasing alkali oxide content, any data which vary monotonically with alkali oxide concentration also vary monotonically with connectivity number.

Finally, structures are often discussed in terms of the fraction of the total building blocks represented by a particular type of unit. For example, we might wish to determine the fraction of the silicon–oxygen tetrahedra which contain one NBO per tetrahedron, *i.e.*, the fraction of Q_3 units, f_3. In this case, using Equations 5.6 and 5.7, we can calculate f_3 from the expression

Table 5.3 *Characteristics of* Q_n *units*

Characteristic	Q_4	Q_3	Q_2	Q_1	Q_0
BO/tetrahedron	4	3	2	1	0
NBO/tetrahedron	0	1	2	3	4
Oxygens/tetrahedron	2.0	2.5	3.0	3.5	4.0
Bridges/tetrahedron	4	3	2	1	0

$$f_3 = \frac{Q_3}{Q_3 + Q_4} = \frac{2x}{100 - x} \tag{5.12}$$

The equations given above provide examples of quantitative information which might be obtained from a simple structural model. The various equations are often used to predict trends in properties or in the results of spectral studies. A number of other useful relationships can be derived as needed once one understands the basic principles used to derive these expressions. A few useful characteristics of the various Q_n units are summarized in Table 5.3 as an aid in performing these calculations.

The expressions for Q_4 and Q_3 derived here only apply in the region from $x = 0$ ($Q_4 = 100\%$) to $x = 33.3$ mol % ($Q_3 = 100\%$). If we continue to add alkali oxides to silica, forming more NBO, we must derive a new set of expressions for the next compositional region, using Q_3 and Q_2 units as building blocks for the network. Once the Q_3 units disappear, we will enter a compositional region where the model predicts only Q_2 and Q_1 units, followed by a region of Q_1 and Q_0 units, and, eventually a region containing only Q_0 units. The compositional regions and appropriate equations for the concentrations of each type of unit within those regions are listed in Table 5.4. Derivation of these equations follows exactly the same procedures as used in the region between 0 and 33.3 mol % alkali oxide. Expressions describing the compositional dependence of one of the network units present, expressed as a fraction of the total number of network units, are also given in this table. Since we have assumed that only two types of network units exist for any composition, the fraction of the network units represented by the second network unit is given by 1 minus the fraction of the unit listed.

The fractional equations given in Table 5.4 can be used to produce a figure (Figure 5.3) showing the theoretical structure of any alkali silicate glass in terms of the concentrations of various Q_n units. The solid lines in this figure represent the fractional concentrations determined using our assumptions. Measurements using NMR indicate that our assumptions

Table 5.4 *Mathematical expressions for glass structural models for binary systems*

System	*Composition Region* (mol % R_2O)	*Equations*	*Fractional equations*
Alkali Silicate	0–33.3	$Q_4 = 100 - 3x$ $Q_3 = 2x$	$f_3 = \frac{2x}{100 - x}$
	33.3–50.0	$Q_3 = 200 - 4x$ $Q_2 = 3x - 100$	$f_2 = \frac{3x - 100}{100 - x}$
	50.0–60.0	$Q_2 = 300 - 5x$ $Q_1 = 4x - 200$	$f_1 = \frac{4x - 200}{100 - x}$
	60.0–66.7	$Q_1 = 400 - 6x$ $Q_0 = 5x - 300$	$f_0 = \frac{5x - 300}{100 - x}$
Alkali borate	0–30.0	$N_3 = 200 - 4x$ $N_4 = 2x$	$N_4 = \frac{x}{100 - x}$
Alkali germanate	0–?	$N_4 = 100 - 2x$ $N_6 = x$	$N_6 = \frac{x}{100 - x}$

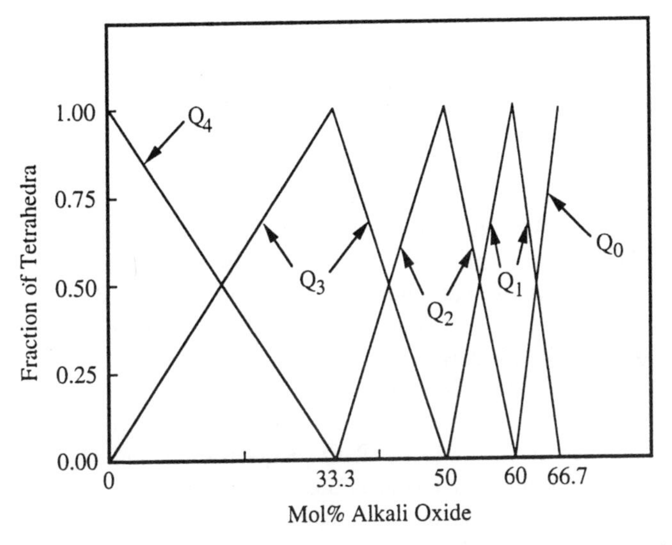

Figure 5.3 *Effect of alkali oxide concentration on the relative theoretical concentrations of* Q_n *units in alkali silicate glasses*

are perhaps a bit too conservative and that more than two units can co-exist in a given glass. Direct measurements of the total NBO content of alkali silicate glasses using ESCA indicates that these equations are quite accurate in predicting the overall concentration of BO and NBO in alkali silicate glasses.

A few, minor additional points can be made regarding the structure of these glasses. There is some evidence that the alkali ions are not randomly distributed throughout the network, but rather exhibit some degree of clustering, even in glasses which are not phase separated. At the very least, the alkali ions probably occur in pairs near the same NBO, as is often indicated in sketches of proposed structures for these glasses (Figure 5.2). There is some speculation that the alkali ions occur in regions which provide pathways *via* alkali interstices through these glasses, with other regions which are rich in silica and are separated by these river-like alkali rich regions. Other studies have been interpreted as indicating the existence of chains or layers of silica rich regions, with alkali ions concentrated between the chains or layers. A number of ring structures have also been proposed as intermediate range order units. At present, while there is certainly evidence to support such models, the existence of intermediate range units in silicate glasses should be considered to be speculative until methods for directly detecting, rather than simply inferring, the presence of such units are developed.

Alkali/Alkaline Earth Silicate Glasses

Ternary glasses containing alkaline earth oxides in combination with silica and alkali oxides, which are commonly called *soda–lime–silica* glasses, usually contain 10–20 mol % alkali oxide, primarily in the form of Na_2O, or soda, 5–15 mol % alkaline earths, primarily as CaO, or lime, and 70–75 mol % silica. In many cases, some of the soda is replaced by K_2O or, less commonly, by Li_2O. Use of dolomite as a source of CaO often means that considerable MgO is also present in the glass. Replacement of CaO and/or MgO by SrO or BaO occurs occasionally in the production of glasses for special purposes other than simply containers or windows.

A simple model for alkali–alkaline earth–silica glasses strongly resembles that proposed for the alkali silicate glasses. If we now write the general composition in the form $x\mathrm{R_2O}$–$y\mathrm{RO}$–$(100-x-y)\mathrm{SiO_2}$, we can generate equations for the concentration of Q_n species analogous to those listed in Table 5.4 by replacing the quantity x by the quantity $(x + y)$. A schematic drawing for a structure of these glasses (Figure 5.2) includes both R^+ and R^{2+} ions, with the necessary condition that every R^+ ion must have a neighboring NBO, while every R^{2+} ion must have two neighboring NBO. This structure provides stronger network linkages at the alkaline earth sites and, since the R^{2+} ions are much more tightly bonded into the structure, sites where the modifier ions are relatively immobile. The replacement of the more mobile alkali ions by the less mobile divalent alkaline earth ions reduces the net mobility of modifier ions through the structure, improving the chemical durability and reducing the ionic contribution to the electrical conductivity of the glass.

Alkali and Alkaline Earth Aluminosilicate Glasses

Aluminum and gallium ions routinely occur in both tetrahedral and octahedral coordination in crystalline materials. Since glasses containing gallium are assumed to be isostructural with those containing aluminum, the following comments regarding aluminosilicate glasses should be viewed as equally applicable to galliosilicate glasses.

It is generally assumed that most, if not all, of the aluminum in these glasses will occur in aluminum–oxygen tetrahedra so long as the total concentration of alkali and/or alkaline earth oxides equals or exceeds that of alumina. These tetrahedra substitute directly into the network for silicon–oxygen tetrahedra. It follows that alumina, which does not readily form a glass by itself, can, however, easily replace silica in the vitreous network. Oxides which act in this manner are said to be *intermediate* in behavior between glass formers and modifier oxides.

Since aluminum–oxygen tetrahedra with four bridging oxygens have an excess negative charge of -1, an associated cation must be present in the vicinity of each such tetrahedron to maintain local charge neutrality. One might then visualize the aluminum–oxygen tetrahedron as a large anion with an effective -1 charge distributed over the entire anion. The associated modifier cation can be located anywhere in the immediate vicinity of this anion. Charge compensation via an alkaline earth ion, with a charge of +2, requires that two aluminum–oxygen tetrahedra occupy nearby sites so that a single associated cation can simultaneously charge compensate two tetrahedra.

Since aluminum oxide only provides 1.5 oxygens per aluminum–oxygen tetrahedron, the oxygen provided by the alkali or alkaline earth oxide is needed to complete the requirement of 2.0 oxygens per tetrahedron for fully linked tetrahedra, *i.e.*, Q_4 species. Since the oxygen supplied by the R_2O and RO components are consumed in the formation of the aluminum–oxygen tetrahedra, they are not available for the formation of NBO. It follows that each added aluminum ion can be considered to remove one NBO from the structure. If the composition of the glass is such that the total concentration of modifier oxides exactly equals that of alumina, the structure should be a fully linked network of Q_4 units, where the cation in any specific Q_4 unit can be either silicon or aluminum, with no NBO present.

This simple model for alkali–alkaline earth aluminosilicate glasses cannot be extended to compositions containing more alumina than total modifier oxides, since an insufficient number of associated cations for charge compensation occurs for such compositions. Various models for the structure of such glasses have been proposed. The two most commonly discussed models suggest that either (a) excess aluminum ions occur in octahedral coordination, with three BO and three NBO in each octahedra, or (b) triclusters of aluminum–oxygen and silicon–oxygen tetrahedra occur, with three-coordinated oxygens connecting the corners of three tetrahedra. If the tricluster contains one aluminum–oxygen tetrahedron and two silicon–oxygen tetrahedra, the overall unit will be charge neutral. If, however, the unit contains two aluminum–oxygen tetrahedra and only one silicon–oxygen tetrahedron, the unit will have a net charge of -1 and will require an associated modifier cation for charge neutrality. Neither of these models has been strongly supported by experimental studies, which have failed to detect the presence of either three-coordinated oxygens or of substantial quantities of octahedrally coordinated aluminum unless the alumina to modifier oxide ratio far exceeds unity. The question of the environment of aluminum ions in glasses containing an excess of alumina over modifier oxides thus still must be considered as unanswered at this time.

Although we cannot adequately describe the structure of all alkali–alkaline earth aluminosilicate glasses, we can deal with the structures of those glasses where the alumina to modifier oxide concentration ratio is less than one. In this case, we can calculate the relative concentrations of Q_n units just as we have for the other systems discussed thus far. Consider, for example, a general composition represented by $(20-x)R_2O–xAl_2O_3–80SiO_2$, where R is any alkali (we could as easily replace the R_2O by RO, where R is now any alkaline earth). If the glass is free of alumina ($x = 0$), this glass will contain Q_4 and Q_3 groups. Addition of alumina will result in the replacement of Q_3 groups by Q_4 groups containing aluminum rather than silicon. We can thus set up an expression for the oxygen balance in the same way as earlier, so that

$$(20 - x) + 3x + 80(2) = 2x(2.0) + (80 - Q_3)(2.0) + Q_3(2.5) \quad (5.13)$$

The terms on the right side of this equation represent the number of oxygens in the aluminum Q_4, silicon Q_4, and silicon Q_3 units, respectively. Solving this equation, we find that the number of Q_3 units is given by the expression

$$Q_3 = 40 - 4x \tag{5.14}$$

This equation thus predicts that the concentration of Q_3 units will be zero if $x = 10$, which is consistent with our contention that the network will be free of NBO when the concentration of modifier oxides exactly equals that of the alumina. The connectivity number of this glass will be 4.0 since all of the tetrahedra are fully linked Q_4 species. Similar calculations can be made for any glass in the ternary so long as the composition contains more modifier oxide than alumina. One must remember, however, to use the appropriate Q_n units required by the alumina-free glass (see Table 5.4).

One other factor must be included in models of networks containing two different Q_4 species. Do the two types of units tend to cluster into regions containing only one or the other species, or do they occur in a random dispersion through the network? Structures of aluminosilicate minerals are often characterized by a dispersion of the aluminum–oxygen tetrahedra in such a manner that no two adjacent tetrahedra contain aluminum ions, *i.e.*, an aluminum–oxygen tetrahedron tends to be connected to four silicon–oxygen tetrahedra. Since the aluminum-oxygen tetrahedra appear to avoid direct links to other such tetrahedra, this arrangement is described by the so-called *aluminum avoidance principle.* Although initially formulated to describe aluminosilicate structures, this principle is often extended to models for other networks containing two different network cations.

Rare Earth Alumino/Galliosilicate Glasses

The structures of these glasses are not well defined. MAS-NMR and Raman studies indicate that the aluminum and gallium ions primarily occupy tetrahedral sites, with lesser concentrations of these ions in five- and six-fold coordination, while the other trivalent cations occupy octahedral sites. The silicon tetrahedra appear to be broadly distributed among Q_4, Q_3, Q_2 and Q_1 species.

Lead Silicate Glasses

The high concentration of lead oxide found in many glasses strongly suggests that PbO does not act as a normal modifier oxide in the structure. Since glasses can be made in compositional regions where our model would indicate that only Q_1 and Q_0 silicon–oxygen tetrahedra exist, and hence no continuous network would occur, the actual network must be somehow linked through the Pb^{2+} ions. Various models for the lead sites suggest that the Pb^{2+} ions may occupy the vertices of PbO_4 pyramids, which are linked to other such pyramids or to silicon–oxygen tetrahedra. Since the Pb–O bonds are relatively weak due to the low field strength of the very large Pb^{2+} ions, the network can be easily disrupted, leading to the low glass transformation temperatures of these glasses.

Lead Halosilicate Glasses

It is possible to replace much of the oxygen by any of the four common halides in glasses containing large lead oxide concentrations. These glasses are anionic electrical conductors, whereas almost all other oxide glasses are cationic electrical conductors. The electrical conductivity of the base lead silicates glasses can be increased by up to five orders of magnitude by the replacement of 25 mol % PbO by PbX_2, where X is F, Cl, Br, or I.

Fluorine ions, which are similar in diameter to oxygen ions, are believed to substitute directly for NBO on the silicon–oxygen tetrahedra, providing breaks in the vitreous network. In contrast, it has been suggested that chlorine, bromine, and iodine replace BO at low halide concentrations and only replace NBO when the concentration of PbX_2 exceeds 5–10 mol %.

STRUCTURAL MODELS FOR BORATE GLASSES

Vitreous Boric Oxide

The current model for the structure of vitreous boric oxide differs significantly from that for vitreous silica. Although boron occurs in both triangular and tetrahedral coordination in crystalline compounds, it is believed to occur only in the triangular state in vitreous boric oxide. All such triangles are connected by BO at all three corners to form a completely linked network. However, since the basic building block of this network is planar rather than three-dimensional, the three-dimensional linkage which occurs in a network of tetrahedra does not exist for vitreous boric oxide. A three-dimensional structure is developed by 'crumpling' of the network, in much the same way that a two dimensional drawing on a sheet of paper develops a third dimension when the paper is crumpled into a ball. Since the primary bonds exist only within

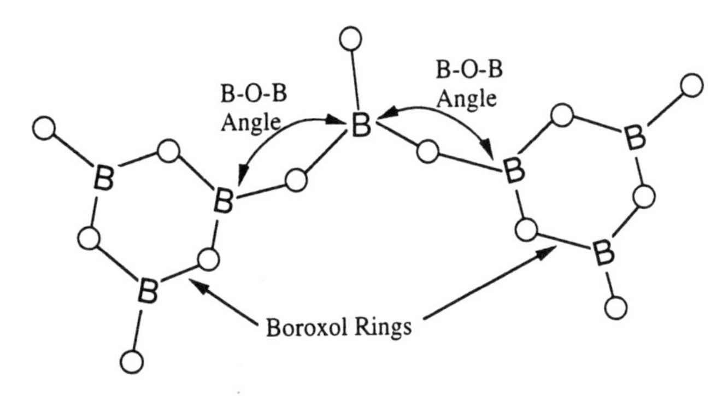

Figure 5.4 *Boroxol ring structures in vitreous boric oxide and alkali borate glasses*

the plane of the paper, bonds in a third dimension (van der Waals bonds in this case) are very weak and the structure is easily disrupted. One result of this structure, for example, can be found in the glass transformation temperature of vitreous boric oxide, which is only 260 °C, as opposed to the T_g of vitreous silica, which is 1100 °C.

The structure of vitreous boric oxide is also believed to contain a large concentration of an *intermediate unit* consisting of three boron–oxygen triangles joined to form a structure known as a *boroxol ring* or boroxol group. These well-defined units are connected by oxygens so that the B–O–B angle is variable and twisting out of the plane of the boroxol group can occur (see Figure 5.4). Arguments supporting the existence of boroxol groups are primarily based on the observation of a very sharp band at 808 cm^{-1} in the Raman spectra of vitreous boric oxide.

Alkali Borate Glasses

Addition of alkali oxides to vitreous silica results in the formation of NBO. Examination of property trends for alkali silicate *versus* alkali borate glasses, however, suggests that this is not the case for alkali borate glasses. Small additions of alkali oxide to silica cause a decrease in T_g, while similar additions to boric oxide cause an increase in T_g. Conversely, small additions of alkali oxides to silica cause an increase in the thermal expansion coefficient, while similar additions to boric oxide cause a decrease in the thermal expansion coefficient. Any potential structural model for alkali borate glasses must directly address this extreme difference in behavior.

If the effects of alkali oxide additions to boric oxide cannot be explained on the basis of NBO formation, how can they be explained? As noted earlier, boron is found in both three- and four-fold coordination in oxide crystals. Perhaps the addition of alkali oxide forces some of the boron to change from triangular to tetrahedral coordination, with no NBO formation. Such a change would actually increase the connectivity of the network, increasing T_g and decreasing the thermal expansion coefficient of the glass, which is consistent with experimental observations. Formation of two boron–oxygen tetrahedra would consume the one additional oxygen provided by the R_2O. Since each tetrahedron would be charge deficient by -1 units, the two alkali oxides would provide sufficient charge compensation for both tetrahedra, in much the same manner as discussed earlier for alkali aluminosilicate glasses. The large $(BO_{4/2})^-$ units now act as anions with a loosely associated alkali cation. A continued increase in the alkali oxide concentration would result in further shift of borons from three- to four-fold coordination.

An interesting phenomenon occurs as even more alkali oxide is added to boric oxide. Many of the property/composition trends reverse directions, with, for example, a minimum in thermal expansion coefficient and a maximum in T_g occurring at higher alkali oxide concentrations. Since such behavior was not observed for the alkali silicate glasses which had been the subject of earlier studies, this behavior was considered to be anomalous for these glasses and hence termed the *borate anomaly*. This anomaly was originally explained by assuming that the conversion of boron from three- to four-fold coordination only occurs until the network reaches some critical concentration of tetrahedrally coordinated boron (originally believed to occur at 16 mol % R_2O), after which additional alkali oxide causes the formation of NBO, which results in a reversal in property/composition trends. This model was widely accepted as the definitive explanation for the borate anomaly for many years.

Eventually, a number of new observations raised questions regarding the original explanation for the borate anomaly. First, and most importantly, NMR was used for the direct measurement of the concentrations of three- and four-fold borons in a large number of alkali borate glasses. The results of these studies clearly demonstrate that the coordination conversion process does occur as proposed, but also show that the maximum in the concentration of tetrahedrally coordinated borons, designated as N_4, does not occur until the composition contains ≈ 35 to 40 mol % alkali oxide. Since this concentration is well past the composition where the property trend reversals occur, a direct correlation between the borate anomaly and the value of N_4 seems doubtful.

More recent property measurements also raised new questions regarding the borate anomaly. Measurements of T_g reveal that the maximum occurs at 27 mol % R_2O, while the minimum in the thermal expansion coefficient for the identical samples occurs at 20 mol % R_2O. Examination of trends for other properties reveals that the borate anomaly occurs at different alkali oxide concentrations for different properties, which is not consistent with the original simple model.

If the original model for the structure of alkali borate glasses is not correct, what are the structures of these glasses really like? Apparently, the original concept of a three- to four-fold boron coordination change, followed at higher alkali oxide concentrations by the formation of NBO is, in fact, correct. This model, however, is not detailed enough to characterize these structures fully. In addition to the boroxol rings found for vitreous boric oxide, alkali borate glasses contain a number of other intermediate structural groups, which are designated by the names of the crystalline compounds in which they occur. The first of these units to be formed upon addition of alkali oxide to boric oxide, for example, consists of a boroxol ring in which one triangle has been converted to a tetrahedron. The replacement of boroxol rings by this unit, designated a tetraborate unit, gives rise to the gradual reduction in the intensity of the sharp Raman band at 808 cm^{-1} and the concurrent growth of another sharp Raman band at $\approx$770 cm^{-1}. Further additions of alkali oxide eventually result in the complete disappearance of boroxol rings and the conversion of tetraborate groups into diborate groups, which have two tetrahedra per three membered ring. Finally, additions of more than $\approx$25 mol % R_2O eventually begin to cause the disruption of the structure and the formation of NBO.

The relative concentrations of these groups are functions of the overall glass composition, as shown in Figure 5.5, which represents a simplified version of the detailed model for these glasses. Although the complete model for these glasses contains a number of other complex intermediate units, this figure is adequate to explain the complex behavior summarized under the general term borate anomaly. The maximum in T_g, for example, is due to the competing effects of the maximum at about 35 mol % R_2O in the concentration of tetrahedra, which increase the connectivity of the network, and the formation of NBO, which begins at $\approx$25 mol % R_2O, which decreases the connectivity of the network. The relationship between the thermal expansion coefficient and the concentrations of the various units has not been established, but it is interesting to note that the minimum in thermal expansion coefficient occurs just as the boroxol rings disappear, the concentrations of tetraborate units passes through a maximum, and where diborate units begin to form. The properties of the alkali borate glasses appear to be strongly connected to the concentrations of the particular intermediate units present in a given glass, while the properties of the alkali silicate glasses

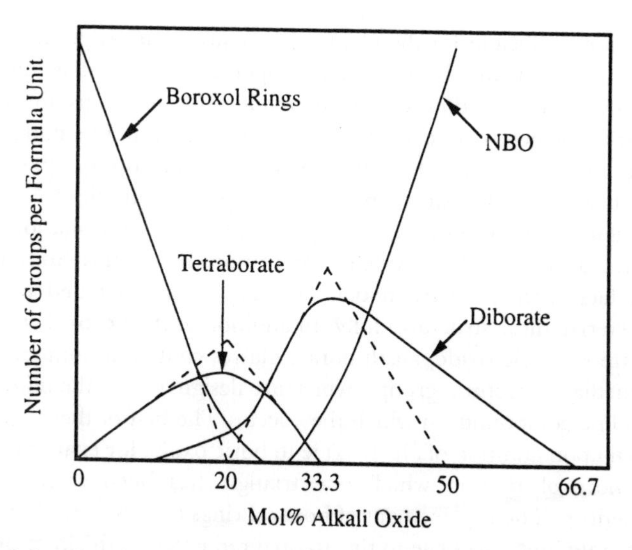

Figure 5.5 *Effect of alkali oxide concentration on the relative concentrations of intermediate range units in alkali borate glasses. Dashed lines indicate simple theory. Solid lines represent experimental results*

are much more simply determined by the relative concentrations of BO and NBO in the network.

Since the structures of the alkali borate glasses only contain Q_4 and Q_3 (triangles which contain only BO) groups as short range order units at small alkali oxide contents, the calculation of the concentrations of these two species is relatively straightforward. Using the same method as before, with a general composition of $xR_2O-(100-x)B_2O_3$, we can state that

$$x + 3(100 - x) = 2Q_4(2.0) + 2[100 - x) - Q_4](1.5) \tag{5.15}$$

Solving for Q_4, we find that $Q_4 = x$ and hence $Q_3 = 100 - 2x$. In this particular case, the structure is usually discussed in terms of the fraction of network which consists of Q_4 units, or $Q_4/(Q_3 + Q_4)$. This fraction, designated as N_4, is then equal to

$$N_4 = \frac{x}{100 - x} \tag{5.16}$$

The connectivity number of the alkali borate glasses in this compositional region is determined by the sum of the contributions of the Q_3 and Q_4 groups. If we consider a glass containing 10 mol % alkali oxide, Equation 5.16 predicts that $N_4 = 0.11$ and thus $N_3 = 0.89$, where N_3 is the fraction of borons in three-fold units. If we multiply the fraction of each type of unit by the number of bridges per unit and sum for all possible units, we obtain a connectivity number of 3.11 for this glass, or 3.11 = [(4)(0.11) + (3)(0.89)].

Data for glasses in alkali borate systems are often plotted *versus* the quantity N_4, or $x/(100-x)$ instead of the more traditional plot against the actual concentration of either R_2O or B_2O_3. The relation expressed by Equation 5.16 is assumed to hold until $N_4 = 0.5$, which occurs at $x =$ 33.3 mol % R_2O. No NBO are assumed to exist in glasses containing $\leq$33.3 mol % R_2O. A glass containing 33.3 mol % R_2O would have a connectivity number of 3.50, which is the maximum possible in this system.

An empirical fit to the NMR data has been proposed to describe the value of N_4 for glasses containing >33.3 mol % R_2O. In this compositional region, N_4 is given by

$$N_4 = \frac{300 - 4x}{500 - 5x} \tag{5.17}$$

This equation predicts that a glass will contain zero N_4 groups when the R_2O concentration equals 75 mol %.

Alkali Aluminoborate Glasses

The structures of alkali aluminoborate and alkali gallioborate glasses are not well understood at this time. A simple model would suggest that aluminum–oxygen tetrahedra would form in preference to boron–oxygen tetrahedra. It can be argued that the boron can remain in three-fold coordination when we add alumina to the glass, but that the aluminum ions must be four-fold coordinated and will, therefore, consume the additional oxygens from the alkali oxide in preference to the borons. Formation of boron Q_4 groups will occur only if there is excess oxygen after formation of the aluminum Q_4 groups.

Since no maximum in the aluminum–oxygen tetrahedra concentration should occur with increasing alumina content, this model would predict that the borate anomaly would not occur for a series of glasses containing a one-to-one ratio of R_2O to alumina. No boron Q_4 groups would ever be formed in these glasses since all of the oxygen supplied by the alkali oxide would be used to form aluminum Q_4 groups. Glasses containing less alumina would contain some boron Q_4 groups, which might be expected to exhibit behavior similar to that of alumina-free borate glasses.

This simple model predicts that the borate anomaly would gradually shift to higher R_2O concentrations as the alumina content of the glasses increases. The anomaly should completely disappear for the one-to-one alkali oxide to alumina ratio glasses. Experimental results for these glasses, however, are in direct conflict with this model. The properties of alkali aluminoborate glasses (and their corresponding alkali gallioborate glasses, which are expected to behave in the same fashion) exhibit borate anomalies at approximately the same alkali oxide concentrations, regardless of the alumina concentration. The simple model, which is widely quoted in the literature, thus fails the basic requirement that it explain the experimental data. No model can be accepted as valid which, however logical, fails to provide an explanation for experimental observations.

At present, no sufficient model for the structure of these glasses can be found in the literature. NMR studies have indicated that aluminum ions occur in four-, five-, and six-fold coordination in these glasses, but a detailed structural model which adequately explains all of the experimental results does not exist.

Alkali Borosilicate Glasses

A large number of commercial glasses are based on alkali borosilicate systems, with a majority of these glasses primarily containing soda instead of any of the other alkali oxides. Most of the commercial glasses, while transparent, are actually phase separated with a very fine scale morphology.

Discussion of structural models for the alkali borosilicate glasses must begin with a consideration of the compositional regions of heterogeneous, *i.e.*, phase separated, and homogeneous glasses. The lithium and sodium borosilicate systems contain very large immiscibility regions, while the immiscibility region in the potassium borosilicate system is much smaller. The high critical temperature (between 950 and 1000 °C) in the lithium borosilicate system leads to the formation of coarse microstructures during cooling from the melt, so that many of these glasses are opaque or translucent and thus obviously phase separated. Since the critical temperature in the sodium borosilicate system is significant lower (between 750 and 800 °C), glasses in this system are easily produced with microstructures of such fine scale that they appear transparent to the naked eye and are often, therefore, assumed to be homogeneous. Many of the commercial sodium borosilicate glasses also contain a small quantity of alumina, which further depresses the immiscibility temperature, leading to an even finer scale microstructure.

Even though the microstructure of sodium borosilicate glasses is so fine that regions of a given phase may be only 5–20 nm in diameter, these regions are still much larger than the atomistic scale regions sampled by most spectral measurements. As a result, models based on NMR or Raman measurements, for example, which assume that their samples are homogeneous glasses for compositions lying within the phase separated region must be inherently incorrect. Examination of the tie lines in the sodium borosilicate system reveals that the glasses containing less than 20 mol % soda actually consist of two vitreous phases. One of these phases is a very silica-rich glass, while the other phase contains almost all of the alkali and boric oxides. Spectra obtained for samples of these glasses are simply sums of the spectra of a glass which resembles vitreous silica rich and that of an alkali borate-rich glass. Interpretation of these spectra must be based on the understanding that these glasses are not homogeneous and that the spectra are geometrically weighted averages of the spectra of the two phases present. Since the alkali borate phases present in these glasses contain a high percentage of boron Q_4 groups, NMR studies indicate that initial addition of soda to borosilicate glasses results in the conversion of boron Q_3 groups to boron Q_4 groups, which are linked to other boron groups in much the same way as found for binary alkali borate glasses. This finding is not surprising since the spectra observed really are those of alkali borate glasses as a result of phase separation of the samples.

Studies of glasses containing >20 mol % soda, where the glasses are homogeneous, can be interpreted in much the same manner as those for the alkali borate glasses. The concentration of boron Q_4 groups passes through a maximum and begins to decrease as the soda concentration increases beyond 33.3 mol %. The network structure contains boron Q_3 and Q_4 groups, as well as silicon Q_n groups. The concentration of NBO gradually increases until glass can no longer be formed.

STRUCTURAL MODELS FOR GERMANATE GLASSES

Vitreous Germania

Vitreous germania has a structure very similar to that of vitreous silica, with basic building block of germanium–oxygen Q_4 units. Since the germanium ion is somewhat larger in diameter than the silicon ion, the Ge–O distance is also somewhat greater, with a bond length of $\approx$0.173 nm. The Ge–O–Ge bond angle is smaller than the Si–O–Si bond angle. Gas diffusion studies suggest that the structure of vitreous germania is more compact than that of vitreous silica and therefore the free, or interstitial, volume of vitreous germania is slightly less than that of vitreous silica. Structural defects are more common in vitreous germania than in vitreous silica, with a measurable concentration of Ge–Ge bonds.

Binary Borogermanate and Silicogermanate Glasses

The structures of borogermanate and silicogermanate glasses consist of fully linked networks containing mixtures of germanium Q_4 groups with either boron Q_3 or silicon Q_4 groups, as appropriate. The properties of glasses in these systems vary smoothly from those of one pure glassformer oxide to those of the other. Although these glasses are not phase separated, there is some spectral evidence for intermediate range order groups which contain only one or the other of the glassformer oxides. The connectivity number for borogermanate glasses varies between 3 and 4, while that of silicogermanate glasses is always 4.

Alkali Germanate Glasses

Two completely different structural models are currently proposed for alkali germanate glasses. The traditional model is based on the observation of maxima in the density and refractive index of alkali germanate glasses with increasing alkali oxide concentration. Since the existence of these maxima was believed to indicate behavior similar to that observed for alkali borate glasses, the term *germanate anomaly* was coined to describe these property trends. Furthermore, since germanium in crystalline GeO_2 polymorphs can be found in both tetrahedral and octahedral coordination, it seemed reasonable to propose that additions of alkali oxide cause a shift in germanium coordination from four- to six-fold, without any NBO formation. Each octahedron must be associated with two monovalent cations for charge neutrality. Eventually, the formation of these germanium Q_6 groups will reach a saturation limit and NBO will begin to form, accompanied by the return of germanium to the tetrahedral state. The increased packing efficiency of the octahedral network units should lead to density increases, while the eventual formation of NBO will result in a decrease in packing efficiency and a decrease in density.

A number of spectral techniques have been used to study these glasses, with the conclusion that the coordination change does occur, but with considerable variation in the composition where the Q_6 concentration should reach a maximum. The theoretical value of N_6 (the ratio of Q_6 to total germanium groups) in this region of unknown upper R_2O concentration is given by the quantity $x/(100-x)$, just as is N_4 in the alkali borate systems.

Although widely accepted, this model has a number of flaws. Shelby (1974), for example, notes that the traditional model predicts a continuous increase in the connectivity of the network as Q_6 units replace Q_4 units. The glass transformation temperatures of alkali germanate glasses, however, immediately decrease by over 100 K upon the addition of as little as 1 mol % of alkali oxide to germania. A minimum in T_g occurs at ≈ 2 mol % R_2O, followed by an increase in T_g to about 15 mol % R_2O, where a maximum occurs, after which T_g decreases rapidly. The presence of the minimum in T_g at 2 mol % R_2O is in direct conflict with predictions based on the simple model for these glasses, while the maximum at ≈ 15 mol % R_2O might be explained by the model. Shelby (1974) suggested that the first, or low alkali concentration, anomaly may be due to the formation of NBO for small alkali oxide contents and that the conversion of germanium to Q_6 units may not occur unless some minimum number of NBO exist in the structure.

Recently, Henderson and Fleet and others have raised questions regarding the possibility that no germanium–oxygen octahedra ever form in these glasses and have proposed the *ring model* for germanate glasses. It has been suggested that the spectral results of other studies have been misinterpreted and that the evidence which actually exists simply indicates that lengthening of the Ge–O bonds occurs as alkali oxides are added to the glass. It is proposed that the bond lengthening is not due to Q_6 formation, but rather to distortions of the tetrahedra which occur when small three-membered rings are formed. These three-membered rings create smaller network interstices, which more closely fit the alkali ions than the four- or six-membered rings proposed to exist in vitreous germania and therefore lead to density increases. Some NBO are formed in all compositions, which agrees with the behavior predicted by the trends in T_g and viscosity noted by Shelby (1974). The maxima in properties are due to eventual saturation of the network with three-membered rings, after which their destruction causes a reverse in property trends with further increases in alkali oxide concentrations.

The correct structural model for alkali germanate glasses is not known at this time. It may prove that both models have some validity, with germanium–oxygen octahedra, three-membered rings, and NBO all formed simultaneously in glasses containing less than 15–20 mol % R_2O. Further additions of alkali oxides may result in the reconversion of germanium–oxygen octahedra to tetrahedra and destruction of three-membered rings, with the formation of much larger numbers of NBO.

Alkali Aluminogermanate Glasses

Addition of alumina to alkali germanate glasses results in a shift in the position of the germanate anomaly to higher R_2O concentrations. The anomaly eventually disappears for glasses which contain equimolar concentrations of alkali oxide and alumina. The low alkali anomaly in T_g and viscosity, however, does not disappear regardless of alumina content. While no direct structural studies have been reported, this behavior suggests that alumina either (a) forms tetrahedra in preference to the formation of germanium–oxygen octahedra or (b) forms tetrahedra in preference to three-membered ring formation. In either case, the structural changes with increasing alkali oxide formation which cause the germanate anomaly in alkali germanate glasses are suppressed by the presence of alumina in the glass. The continued existence of the low alkali anomaly implies that the structural characteristics of the network causing this effect are independent of the presence of alumina.

Fluorogermanate Glasses

Alkali germanate glasses retain fluorine in their networks at a much higher level than possible in alkali silicate glasses. Glass formation in alkali germanate systems is actually improved by replacing alkali oxides by alkali fluorides. Property trends for alkali fluorogermanate glasses are similar to those for the alkali germanate glasses, with both low alkali and germanate anomalies in T_g and maxima in density and refractive index with increasing alkali fluoride concentration. Speculations regarding the structures of these glasses are based on the models for the alkali germanate glasses, with fluorine replacing oxygen as a non-bridging anion on a portion of the network units.

STRUCTURAL MODELS FOR PHOSPHATE GLASSES

The structure of vitreous phosphoric oxide is also based on a tetrahedral building block. Since phosphorus is a pentavalent ion, the formation of a phosphorus/oxygen tetrahedron with four bridging oxygens would result in a unit with a net positive charge of +1. A charge balanced tetrahedron can be created, however, if one of the oxygens forms a double bond with the P^{5+} ion, while the other three oxygens form BO with adjacent tetrahedra. The two-dimensional network formed by the connection of these tetrahedra at three corners thus has the same connectivity as that of vitreous boric oxide, even though the building blocks are four-cornered tetrahedra. This network is easily disrupted, resulting in a very low glass transformation temperature for vitreous phosphoric oxide.

Details of the structure of vitreous phosphoric oxide are dependent upon the source of P_2O_5 used to produce the melt. Crystalline phosphoric oxide exists in three polymorphic forms: hexagonal, orthorhombic, and tetragonal. These crystals all contain phosphorus/oxygen tetrahedra, but contain different intermediate range units in the form of rings with different numbers of tetrahedra per ring. Glasses produced using different starting materials retain some of the structural details of the crystalline form for short melting times, with only gradual convergence of properties toward equilibrium values after extended time at high temperature.

Addition of alkali and alkaline earth oxides to phosphoric oxide cause the breaking of rings and the conversion of the network to a system of entangled linear chains of phosphorus/oxygen tetrahedra crosslinked by the monovalent or divalent ions. These chains can be oriented during fiber drawing, producing glasses with directional properties. The structures of these glasses thus resemble those of organic polymeric glasses and provide a bridge between the structures of inorganic and organic glasses. Recent studies, in fact, have shown that organic molecules can be incorporated into phosphate glasses to produce materials with interesting optical properties.

Alkali phosphate melts will react with nitrogen to produce nitrided glasses if nitrogen is present in a highly reactive form, which can either be introduced by use of nitrides in the batch or by exposure of the melt to ammonia, which decomposes to release nitrogen ions. Exposure of melts to relatively inert N_2 molecules results in very little reaction. Nitrogen replaces oxygen in the network, allowing three tetrahedra to share a single corner. Since oxygen ions only allow sharing of two corners of the network units, replacement of oxygen by nitrogen increases the connectivity of the structure, improving the chemical durability and increasing the glass transformation temperature. Nitrided glasses have been produced with durabilities superior to that of commercial soda–lime–silica bottle glasses.

STRUCTURES OF OTHER INORGANIC OXIDE GLASSES

A number of other important inorganic glasses can be formed in systems which do not contain any of the four traditional glassforming oxides. Aluminate glasses are easily formed in a small region of the binary system $CaO–Al_2O_3$ even though alumina itself does not form a glass using traditional melt-cooling methods. These glasses are believed to consist of networks of aluminum–oxygen tetrahedra with Ca^{2+} ions serving as modifiers and to preserve charge balance. There is recent evidence that at least some aluminum–oxygen octahedra, and possibly even some five-fold aluminum–oxygen units, also exist in these glasses. Related glasses can be formed in the PbO–Ga_2O_3 system, with suggested structural models similar to those for the aluminate glasses. The lead gallate glasses are particularly interesting because of their excellent infrared transmission and extremely high refractive index. They also have very high Verdet coefficients for diamagnetic glasses as a result of their high lead contents.

Tellurium oxide acts as a glassformer in the presence of a few mol % of a large number of other oxides, including the alkali and alkaline earth oxides and the oxides of a large number of tri-, tetra-, and pentavalent cations. Although little structural evidence exists, these glasses were originally believed to contain networks of tellurium–oxygen octahedra. More recent models suggest that the glass is best described as consisting of chains of three- or four-fold coordinated polyhedra formed by distortion of trigonal bipyramids. All of these models are quite speculative. Tellurium oxide is so reactive with other oxides that melting in silica or alumina crucibles will produce glasses which are highly contaminated with dissolved materials from the melting crucibles.

Glasses can be produced in many other oxide systems, including, but not limited to, titanates, vanadates, arsenates, antimonates, bismuthates, tungstates, molybdenates, and in oxide systems containing more complex anions such as carbonates, nitrates, and sulfates. Incorporation of some carbonate into alkali borate glasses at very high alkali oxide concentrations to form mixed oxide–carbonate glasses has also been observed. Although the properties of many of these glasses have been widely studied, structural models are very speculative at present.

HALIDE GLASSES

Fluoroberyllates

Beryllium fluoride melts to form a very viscous liquid. Crystalline BeF_2 contains beryllium–fluorine tetrahedra in quartz and cristobalite forms, *i.e.*, the crystalline forms of BeF_2 closely resemble those of SiO_2. Be^{2+} ions are almost identical in size with Si^{4+} ions, while F^- ions are almost identical in size with O^{2-} ions. It is not surprising to find that vitreous beryllium fluoride closely resembles vitreous silica, with a network made up of tetrahedra connected at all four corners. The similarity between these glasses is so striking that beryllium fluoride has been called a *weakened model* for silica.

Clear glasses can be formed in a number of binary systems using BeF_2 and either alkali or alkaline earth fluorides. Since these glasses do not display any visual signs of phase separation, they were originally considered to be homogeneous. Structural models were proposed which are directly analogous to those for alkali silicate glasses, with the replacement of non-bridging oxygens by *non-bridging fluorines*, or *NBF*. Later studies using transmission electron microscopy, which can detect phase separation with a much finer scale than that obvious to the naked eye, reveal that essentially all of these glasses are phase separated, with BeF_2-rich and modifier fluoride-rich phases. If these glasses are phase separated over most of the glassforming region, then discussions of their structures in terms of models for homogeneous glasses is superfluous.

Glasses Based on $ZnCl_2$

Zinc chloride has only a modest viscosity at the melting temperature, so it does not form a glass as readily as BeF_2. $ZnCl_2$ glass is also easily dissolved in water, so it is of little practical use. Addition of alkali halides improves the glassforming behavior and allows formation of larger samples. The network is believed to consist of zinc–chlorine tetrahedra linked at the corners and to resemble those of BeF_2 and SiO_2. Little is known regarding the effect of modifiers on the network of these glasses.

Fluorozirconate (Fluorohafnate) Glasses

The basic building blocks of these highly ionic glasses are less well defined than are those determined by the more covalent bonding found within silicon–oxygen tetrahedra, where the sp^3 bonding tends to yield very well defined tetrahedra. Various reports have proposed that Zr^{4+} or Hf^{4+} ions are coordinated by six, seven, or eight F^- ions. Structures may contain a mixture of coordination units, such that the average coordination number ranges from six to eight. These units may be connected by corners to form either chains, similar to those found in phosphate glasses, or three-dimensional networks. In contrast to oxide glasses, both edge- and face-sharing of polyhedra have also been suggested in proposed structures of fluorozirconate glasses. Non-bridging fluorines are also believed to exist in these glasses. Modifier cations such as Ba^{2+} and other monovalent and divalent cations are usually present in neighboring interstices to provide charge neutrality. The role of trivalent cations, which are usually present in these glasses, is even less well defined, but they probably occupy higher coordination states (coordination numbers >4) and substitute for Zr^{4+} in the network.

CHALCOGENIDE GLASSES

The term *chalcogenide* refers to the elements in the group VIA column of the periodic table, *i.e.*, O, S, Se, Te, and Po. Since the isotopes of polonium are radioactive, no significant studies have been reported using polonium in glasses. If we consider the other four chalcogenides, we find that oxygen differs from the others in that (a) it is a gas at room temperature and (b) oxygen exists as individual O_2 molecules, while S and Se melts contain rings and chains which can interact to form a vitreous structure. Melts of tellurium, which is much more difficult to form as a glass, also contain chains, with a structure similar to that of molten Se. It follows that the term *chalcogenide glasses* is limited to compositions containing S, Se, and/or Te. While many other elements can be added to these glasses, the primary chalcogenide glasses contain one or more of these three elements along with elements from groups IVA (Si, Ge, Sn) and VA (P, As, Sb, Bi). If the glasses also contain a halide, they are commonly termed *chalcohalide glasses*.

Structural models for these glasses are based on the high degree of covalent bonding between chalcogenide atoms. Sulfur melts contain eight-membered rings at low temperatures, with covalent bonding between adjacent sulfur atoms. These rings begin to convert to extremely long chains of S atoms at temperatures above 160 °C, with chain lengths exceeding 10^6 atoms. Molten Se contains somewhat shorter chains, while molten Te consists of even shorter chains. These chains thus form polymeric structures similar to those found in some phosphate systems and common in organic glasses.

Addition of elements from groups IVA and VA of the periodic table results in *crosslinking* of these chains (Figure 5.6) to form more highly linked structures. The coordination number of elements in these glasses is given by the *8−N rule*, which states that the coordination number is given by 8 minus the number of outer electrons in the element, *i.e.*, the column number (halides are not considered when using this rule). Elements from group VA (P, As, Sb) thus are expected to occur in three-fold coordination, while elements from group IVA (Si, Ge, Sn) are expected to occur in four-fold coordination. The average coordination number is given by the weighted coordination of the component atoms, so that a As_2Se_3 glass, for example, will have an average coordination number of 2.4. It follows that the connectivity number of these glasses is a strong function of the composition, increasing from 2.0 for vitreous S

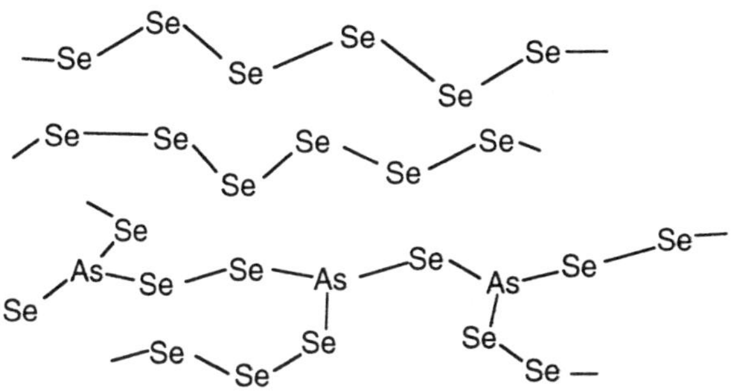

Figure 5.6 *Schematic representation of the structure of vitreous selenium and of an arsenic selenide glass*

and Se as other elements are added to the composition. Addition of halides, on the other hand, forms non-bridging species, which terminate chains and reduce the connectivity number of the glass.

Two models exists for prediction of the number of bonds between different elements. If a glass contains two elements, A and B, we might find bonds between two A atoms, between two B atoms, or between one A and one B atom. If the probability of each type of bond is given by the statistical probability based on the ratio of each type of atom in the composition, we term the structure a *random covalent network*, or *RCN*. On the other hand, if we assume that A–B bonds are always favored and that homopolar bonds between two A or two B atoms will only occur when forced by the deviation of the stoichiometry from equimolar concentrations of A and B, we term the structure a *chemically ordered covalent network*, or *COCN*.

ORGANIC GLASSES

Organic glasses consist of carbon–carbon chains which are so entangled that rapid cooling of the melt prevents reorientation into crystalline regions. These structures closely resemble those of vitreous sulfur and selenium, which also consist of entangled chains. The chains in organic glasses can also be crosslinked, just as they are in chalcogenide glasses, with consequent changes in their properties. Increasing the degree of crosslinking, for example, increases the viscosity of the melt and the glass transformation temperature. In general, the properties of organic glasses closely parallel those of the inorganic glasses with chain-based structures, including the ability to produce materials with oriented properties by application of stress during forming.

Small regions of oriented chains often exist in organic glasses, so that many of these materials actually resemble low-crystallinity glass-ceramics. Proper heat treatments can increase the crystallinity of many of these glasses. Properties of the partially crystalline materials follow the same trends with increasing crystallinity as observed for inorganic glass-ceramics.

AMORPHOUS METALS

Alloys containing both a metal and a metalloid (P, Si, Ge, B) can often be rapidly cooled directly from a melt to form a material with an amorphous structure and which exhibits a glass transformation temperature when heated. Many additional *amorphous metals* can be formed by deposition from the vapor phase. With few exceptions, these glasses can only be obtained as thin ribbons or films as a result of the rapid cooling rate required to prevent crystallization. Common examples of these glasses include $Pd_{80}Si_{20}$ and $Ni_{80}P_{20}$, which are often discussed in basic studies, and $Fe_{40}Ni_{40}P_{14}B_6$, which is a commercial material sold in ribbon form.

Structural models for these materials include variations of the random network theory, crystallite theory, and a *dense random packing* of spheres. A random packing of monosized spheres contains many interstices which are too small to contain spheres of that size, but which could contain smaller spheres. As a result, the density of amorphous metals predicted by the random packing model is somewhat less than that of a close packed structure of the same spheres. The metalloid atoms, which occupy the interstices, interfere with the reorganization of the melt into the crystalline structure during cooling and thus radically improve the glassforming ability as compared to the metalloid-free melt.

Chapter 6

Viscosity of Glassforming Melts

INTRODUCTION

The kinetic model of glass formation indicates that the temperature dependence of the viscosity plays a major role in determining the ease of glass formation for any melt. Glasses are most easily formed if either (a) the viscosity is very high at the melting temperature of the crystalline phase which would form from the melt, or (b) the viscosity increases very rapidly with decreasing temperature. In either case, crystallization is impeded by the kinetic barrier to atomic rearrangement which results from a high viscosity.

In addition to controlling the ease of glass formation, viscosity is also very important in determining the melting conditions necessary to form a bubble-free, homogeneous melt, the temperature of annealing to remove internal stresses, and the temperature range used to form commercial products. The viscosity also determines the upper use temperature of any glass object and the conditions under which *devitrification* (crystallization) may occur. The very high viscosity encountered in the glass transformation range leads to *viscoelastic* behavior, and to time dependence in many of the properties of the melt.

VISCOSITY DEFINITIONS AND TERMINOLOGY

Viscosity is a measure of the resistance of a liquid to shear deformation, *i.e.*, a measure of the ratio between the applied shearing force and the rate of flow of the liquid. If a tangential force difference, F, is applied to two parallel planes of area, A, which are separated by a distance, d, the viscosity, h, is given by the expression

$$\eta = \frac{Fd}{Av} \tag{6.1}$$

where v is the relative velocity of the two planes. If the velocity varies directly with the applied shear force, the viscosity is independent of force and the liquid is said to behave as a *Newtonian* liquid. At high shear stresses, many glassforming melts exhibit an apparent decrease in viscosity with increasing shear stress. This form of non-Newtonian behavior is known as *pseudoplastic* flow, or *shear thinning*, and is important in high shear rate forming processes.

The original unit for viscosity was based on the cgs system, where the viscosity is given in dyne s cm^{-2}. This unit, which is termed a *poise* and given the symbol P, is used in virtually all literature prior to 1970 and is still used extensively throughout the glass industry. In SI units, which have replaced cgs units in much of the recent literature, viscosity is given in N s m^{-2}, or, since a pascal is a N m^{-2}, the viscosity is reported in Pa s. Since 1 Pa s = 10 P, the conversion of viscosity data from one unit to the other is very straightforward. The viscosity of water at room temperature is $\approx$0.01 P, or 0.001 Pa s.

Fluidity is the reciprocal of the viscosity. A melt with a large fluidity will flow readily, whereas a melt with a large viscosity has a large resistance to flow. While fluidity is often used in dealing with ordinary liquids, virtually all literature dealing with glassforming melts discusses flow behavior in terms of the viscosity.

A number of specific viscosities have been designated as reference points on the viscosity/temperature curve for melts. These particular viscosities have been chosen because of their importance in various aspects of commercial or laboratory processing of glassforming melts. Several other reference temperatures which occur at approximate viscosities are also routinely used by glass technologists. These reference points are summarized in Table 6.1 and are shown on a typical curve of viscosity *versus* temperature for a soda–lime–silica melt in Figure 6.1.

The viscosity of a typical melt under conditions where fining and homogeneity can be obtained in a reasonable time is termed the *melting temperature.* Melting usually occurs at a viscosity of $\leq$10 Pa s for commercial glasses, but can occur at lower viscosities for non-silicate and, particularly, non-oxide glasses. Since this temperature is not truly a melting point in the classic sense, but rather simply a processing temperature, the term *practical melting temperature* should be used to distinguish between the true melting points of crystals and a reference viscosity which is based entirely on pragmatic considerations.

Formation of a glass object from a melt requires shaping a viscous

Table 6.1 *Viscosity reference temperatures*

Name of reference temperature	*Viscosity* (Pa s)
Practical melting temperature	$\approx 1-10$
Working point	10^3
Littleton softening point	$10^{6.6}$
Dilatometric softening temperature	10^8-10^9
Glass transformation temperature	$\approx 10^{11.3}$
Annealing point	10^{12} or $10^{12.4}$
Strain point	$10^{13.5}$

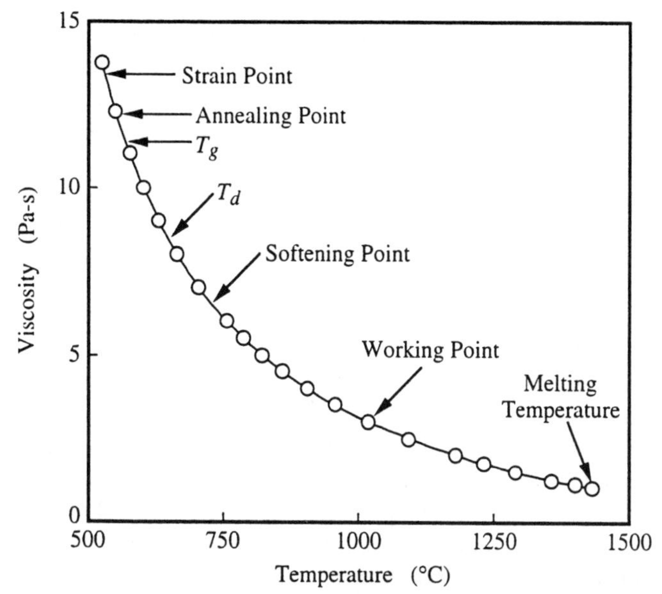

Figure 6.1 *Typical curve for viscosity as a function of temperature for a soda–lime–silica melt (NIST Standard No. 710). Defined viscosity points are indicated on the figure*

mass of liquid, termed a *gob*, by some process involving deformation of the material. The melt must be fluid enough to allow flow under reasonable stresses, but viscous enough to retain its shape after forming. Commercial forming methods require very precise control of the viscosity throughout the forming process in order to achieve high throughput and high yield of acceptable products. Melt is typically delivered to a processing device at a viscosity of 10^3 Pa s, which is known as the *working point.* Once formed, an object must be supported until the viscosity reaches a value sufficiently high to prevent deformation under its own weight, which ceases at a viscosity of $10^{6.6}$ Pa s, which is termed the *softening point.* The temperature range between the working and softening points is known as the *working range.* Melts which have a large working range are often referred to as *long glasses,* while those with a small working range are called *short glasses.* If the working range occurs at high temperatures relative to the working range of typical soda–lime–silica melts, the composition is termed a *hard glass.* On the other hand, if the working range is below that of soda–lime–silica melts, the composition is termed a *soft glass.* This particular terminology is often confusing since the terms hard and soft in this context do not refer to the resistance to scratching usually designated by these same terms.

The softening point is more properly termed the *Littleton softening point,* after the specific test used to define this reference point. The viscosity of $10^{6.6}$ Pa s does not represent the deformation temperature for all objects. This particular reference point is defined in terms of a well-specified test involving a fiber ≈ 0.7 mm in diameter, with a length of 24 cm. The softening point is defined as the temperature at which this fiber elongates at a rate of 1 mm min^{-1} when the top 10 cm of the fiber is heated at a rate of 5 K min^{-1}. In fact, if the density of the fiber is significantly different from that of a typical soda–lime–silica composition, the viscosity will not be exactly $10^{6.6}$ Pa s at this temperature.

Once an object is formed, the internal stresses which result from cooling are usually reduced by *annealing.* The *annealing point* (cited in various sources as either 10^{12} or $10^{12.4}$ Pa s), which is also determined using a fiber elongation test, is defined as the temperature where the stress is substantially relieved in a few minutes. The *strain point* ($10^{13.5}$ Pa s) is defined as the temperature where stress is substantially relieved in several hours. The strain point is determined by extrapolation of data from annealing point studies. Other tests are also used for these two reference points, with slightly different results.

Two other reference temperatures are often quoted for glassforming melts. While neither of these temperatures represent exact viscosities, they are convenient for relative comparison of the viscosity of different compositions. The *glass transformation temperature,* T_g, can be determined from measurements of the temperature dependence of either the heat capacity or the thermal expansion coefficient during reheating of a glass. This temperature is somewhat dependent upon the property measured and on the heating rate and sample size used in the measurement. As a result, different studies will report slightly different values for T_g for supposedly identical glasses. Moynihan has shown that the viscosity corresponding to T_g for common glasses has an average value of

$10^{11.3}$ Pa s. This value appears to decrease for glasses with very low glass transformation temperatures.

Another viscosity point can be obtained from thermal expansion curves. The *dilatometric softening temperature*, T_d, is usually defined as the temperature where the sample reaches a maximum length in a length *versus* temperature curve during heating of a glass. This temperature, which will be discussed in more detail in Chapter 7, varies slightly with the load applied to the sample by the dilatometer mechanism and the sample size. The viscosity corresponding to T_d lies in the range 10^8–10^9 Pa s.

VISCOELASTICITY

At low viscosities, glassforming melts usually behave as Newtonian liquids which immediately relax to relieve an applied stress. At extremely high viscosities, however, these liquids respond to the rapid application of a stress as if they were actually elastic materials. It follows that there must exist an intermediate range of viscosities where the response of these melts to application of a stress is intermediate between the behavior of a pure liquid and that of an elastic solid. Since this behavior has aspects of both viscous flow and elastic response, it is known as *viscoelasticity*, or viscoelastic behavior.

Since the response of a liquid to the application of an external stress is dependent upon the rate of application of that stress, viscoelasticity can occur over a wide range of viscosities. For common rates of stress application, these viscosities lie in the region of the glass transformation range, particularly in the range from 10^{13} to 10^8 Pa s. The most common basic model for viscoelasticity, known as the *Maxwell model*, is shown in Figure 6.2. The sample is considered to consist of an elastic element, represented by the spring, in series with a viscous flow element, represented by the piston in a cylinder filled with viscous liquid. Since the piston/cylinder arrangement is known as a dashpot, models based on these two elements are frequently called spring and dashpot models. The combined spring in series with a dashpot arrangement is known as a *Maxwell element*.

If we consider the application of a force to a Maxwell element, we find that there is an instantaneous displacement of the point A at the bottom of the element to point B. If the viscosity of the liquid is infinite, the displacement is entirely due to stretching of the spring and the response is said to be perfectly elastic. If we now remove the force, the bottom of the element will return to point A.

The opposite response will occur if the viscosity of the liquid is infinitely small. The dashpot will offer no resistance to the displacement,

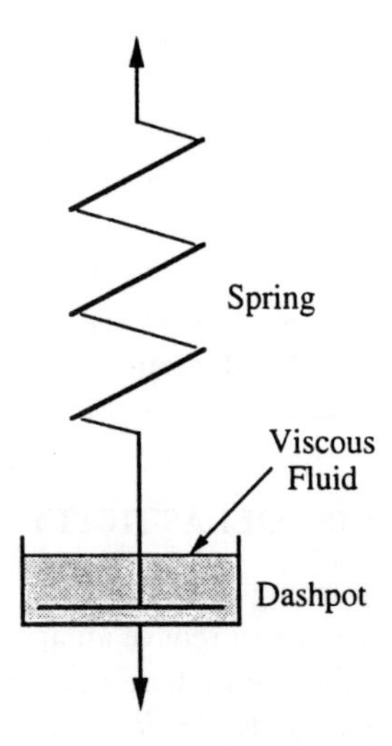

Figure 6.2 *The Maxwell model for relaxation of a viscoelastic material*

the spring will not stretch, and point A will be permanently displaced to point B. There will be no elastic recovery upon removal of the force.

Finally, consider the case where the viscosity of the liquid has an intermediate value. Since flow requires time, the instantaneous response of the element to the displacement will be the same as that for a purely elastic material. The application of the force to the piston, however, will result in flow of the liquid to relieve the strain in the spring and the piston will rise in the dashpot. Eventually, the force will decrease to zero as the spring recovers to its original length and the displacement of the bottom of the element to point B will be permanent. If we calculate the stress as a function of time, we find that

$$\sigma_t = \sigma_0 \exp\left(-\frac{Gt}{\eta}\right) \tag{6.2}$$

where σ_t is the stress at time, t, σ_0 is the stress at time zero, G is the shear modulus, and η is the viscosity of the liquid. The ratio η/G has the dimensions of time and is equal to the time required for the stress to decay to 1/e, or 0.367, of its initial value. This time, which is commonly represented by tau (τ), is known as the *relaxation time*. Replacing η/G by τ, we can thus write the expression

$$\sigma_t = \sigma_0 \exp\left(-\frac{t}{\tau}\right) \tag{6.3}$$

which describes an exponential relaxation curve. If τ is very small, the relaxation will occur so rapidly that normal measurements may indicate that the relaxation process is instantaneous. If τ is very large, the relaxation rate may be so slow that no relaxation is detected in routine measurements.

If we assume that the modulus is essentially temperature-independent, we can ascribe the temperature dependence of τ to the temperature dependence of the viscosity. Assuming a typical room temperature value of G of $\approx 10^{11}$ N m^{-2}, the value of τ at a viscosity of 10^{12} Pa s is 10 seconds. Relaxation of approximately 95% of the initial stress occurs by a time of 3τ, or 30 seconds. It follows that relaxation at temperatures in the glass transformation range would be expected to occur over times ranging from hours to seconds as the viscosity decreases with increasing temperature. Since most experimental measurements of T_g involve heating specimens at rates of 3 to 20 K min^{-1} (0.05 to 0.33 K s^{-1}), we expect that these measurements will indicate relaxation processes occurring on similar time scales, and that T_g values will represent viscosities of 10^{11} to 10^{12} Pa s.

The simple Maxwell model is useful for order of magnitude calculations of relaxation times, but does not adequately represent the behavior of actual glassforming melts. Better models can be obtained by including *Voigt–Kelvin elements*, which consist of a spring and dashpot in parallel, or by coupling a number of Maxwell elements in parallel. A Maxwell element in series with a Voigt–Kelvin element (known as a *Burger element*) may be a better model for actual materials in the glass transformation range. A Burger element actually contains more than one relaxation time. Actual glasses may be best described by multiple Maxwell and Voigt–Kelvin elements, so that a spectrum of relaxation times occurs.

VISCOSITY MEASUREMENT TECHNIQUES

The viscosity of glassforming melts is highly temperature dependent, varying by 12 or more orders of magnitude between the strain point and the practical melting point of a fluid melt. As a result, the measurement of viscosity for a given composition over a wide temperature range requires the use of several different techniques, each of which is restricted to a limited range of viscosity values. Common viscometers are based on direct measurement of the viscosity using a rotation viscometer, the rate of descent of a falling sphere, or the rate of deformation of a plate, fiber, or beam. Less commonly used methods are based on the rate of penetration into the surface of a melt, the torsional deflection of a hollow tube under a torque, or the shearing of a thin disk between a cone and a flat plate.

Rotation Viscometers

Rotation viscometers are commonly used at room temperature to measure the viscosity of a wide variety of liquids in the range $1–10^4$ Pa s. Use of these viscometers at temperatures up to 1600 °C requires that the parts exposed to the melt be constructed of platinum or platinum alloys. These viscometers consist of a small cylinder, or spindle, which is rotated inside a large cylindrical crucible containing the melt. In other versions of this viscometer, the crucible is rotated and the torque exerted on the spindle by the melt is measured. The viscosity range covered by this method can be extended by measuring the time required for the spindle to rotate through a defined angle of deflection ($10^{3.5}–10^{6.5}$ Pa s) or by measuring the torque required to twist the spindle through a small angle ($10^{4.5}–10^9$ Pa s). This method requires use of a few hundred grams of glass to provide a sufficient melt size for reliable measurements.

In the most basic version, the viscosity is determined from measurements of the torque, T, on the spindle and use of Equation 6.4, which relates the viscosity to the torque and the dimensions of the spindle and the cylinder, *i.e.*

$$\eta = \frac{1}{4\pi L}\left(\frac{1}{r^2} - \frac{1}{R^2}\right)\left(\frac{T}{\omega}\right) \tag{6.4}$$

where L and r are the length and radius of the spindle, respectively, R is the inner radius of the cylinder holding the melt, and ω is the angular velocity of the spindle rotation. In normal practice, the actual speed, torque, and dimensions of the instrument components need not be known if the instrument is calibrated at room temperature using oils or other liquids of known viscosity. A calibration constant can then be calculated so that only the ratio of the torque to the rotational velocity need be known to determine the viscosity.

Falling Sphere Viscometers

Viscosities can be measured directly through the determination of the resistance of a liquid to the motion of a sphere falling through the liquid under the influence of gravity. The viscosity is given by the *Stokes equation*

$$\eta = \frac{2}{9}\frac{r^2 g}{v}(\rho_s - \rho_m) \tag{6.5}$$

where r is the radius of the sphere, g is the gravitational constant, v is the velocity of the sphere, and ρ_s and ρ_m represent the densities of the sphere and melt, respectively. A counterbalance is frequently attached to the sphere to reverse its direction of motion, with data collected during both descent and rise of the sphere. This method yields data in the range 1–10^6 Pa s.

Fiber Elongation Viscometers

The most widely used viscometers are based on measurements of the rate of elongation of a fiber of known dimensions under a known load. This method can be used for viscosities ranging from 10^5 to 10^{12} Pa s. This method is also used for the determination of the Littleton softening and annealing reference points. Since the method requires formation of a long fiber for a specimen, it is well suited for many easily worked commercial glasses, but difficulties in the formation of good fibers from many experimental compositions often limit the use of this method for basic research studies.

Fiber elongation measurements are based on the rate of elongation, $\mathrm{d}L/\mathrm{d}t$, where L is the fiber length, of a fiber of cross-sectional area, A, which is suspended vertically in a furnace. The elongation rate is determined by the viscosity of the melt and the applied stress, F/A, where F is the force applied to the fiber. The viscosity is then given by the expression

$$\eta = \frac{LF}{3A(\mathrm{d}L/\mathrm{d}t)} \tag{6.6}$$

Since the area of the fiber is continually decreasing as the fiber elongates, a correction for the changing fiber area must be applied throughout the measurement. The large surface to volume ratio of the fiber also frequently results in compositional changes at the fiber surface either by reaction with atmospheric gases or by evaporation of melt components.

Beam-bending Viscometers

Transformation range viscosities (10^8–10^{13} Pa s) are often measured by the *beam-bending* method, in which a small beam of known cross-sectional area, A, is placed in a three-point bending configuration with a load, M, applied at the center of the beam. The viscosity is given by the expression

$$\eta = \frac{gL^3}{2.4 I_c V}\left(\frac{M + AL\rho}{1.6}\right) \tag{6.7}$$

where L is the length of the specimen between the support spans, I_c is the moment of inertia of the beam, V is the deflection rate of the mid-point of the beam, and ρ is the density of the material. The second term in parentheses, $AL\rho/1.6$, accounts for the contribution of the mass of the beam to the bending load. This term is frequently negligible when compared to the added mass, M, and is often neglected in the calculation, especially for viscometers which use very small samples.

The ease of sample preparation for the beam-bending method makes this technique particularly suitable for research studies. Any beam shape, including rods or tubing in addition to square or rectangular bars, can be used, provided the moment of inertia can be calculated. Viscometers have been designed which use samples as small as 1 × 1 × 10 mm, which allows measurements on compositions which can only be formed as glasses by rapid quenching.

Other Viscometers

A number of other methods are occasionally used for viscosity measurements. The most common are the *parallel plate viscometer*, used in the 10^5–10^8 Pa s range, the *penetration viscometer*, used in the 10^8–10^{12} Pa s range, and the *torsion viscometer*, used in the 10^{11}–10^{14} Pa s range. Although each of these methods has advantages under specific conditions, none has gained wide acceptance in the glass community.

TEMPERATURE DEPENDENCE OF VISCOSITY

Two mathematical expressions, the Arrhenian equation and the Vogel–Fulcher–Tamman equation, are commonly used to express the temperature dependence of the viscosity of glassforming melts. At one extreme, we find that the viscosity can often be fitted, at least over limited temperature ranges, by an Arrhenian expression of the form

$$\eta = \eta_0 \exp(\Delta H_\eta / RT) \tag{6.8}$$

where η_0 is a constant, ΔH_η is the activation energy for viscous flow, R is the gas constant, and T is the temperature in K. In general, Arrhenian behavior is observed within the glass transformation range (10^{13}–10^9 Pa s) and at high temperatures where melts are very fluid. The activation energy for viscous flow is much lower for the fluid melt than for the high viscosity of the transformation region. The temperature dependence between these limiting regions is decidedly non-Arrhenian, with a continually varying value of ΔH_η over this intermediate region.

A relatively good fit to viscosity data over the entire viscosity range is provided by a modification of Equation 6.8 which effectively includes a varying activation energy for viscous flow. This expression was derived independently by several workers and is usually called the *Vogel–Fulcher–Tamman* (or VFT) equation in recognition of each of their contributions. However, since a paper by Fulcher provided most of the early recognition of the utility of this equation, it is also often simply called the *Fulcher equation*, particularly within the glass industry. The VFT equation adds a third fitting variable, T_0, to the Arrhenian expression to account for the variability of the activation energy for viscous flow and replaces the ΔH_η with a less defined variable, B, as indicated by the expression

$$\eta = \eta_0 \exp(B/(T - T_0)) \tag{6.9}$$

This expression is most often written in the form actually used by Fulcher

$$\log \eta = -A + \frac{B \times 10^3}{T - T_0} \tag{6.10}$$

where the constant A replaces η_0 and T and T_0 are given in °C rather than K. The value of T_0 for a given composition is always considerably less than the value of T_g for that composition.

While the VFT equation provides a good fit to viscosity data over a wide temperature range, it should be used with caution for temperatures at the lower end of the transformation region, where ΔH_η becomes constant. The VFT equation always overestimates the viscosity in this temperature regime.

The degree of curvature of viscosity/temperature plots can vary over a wide range due to variations in the value of T_0 relative to T_g. If T_0 is equal to zero, the viscosity/temperature curve will exhibit Arrhenian behavior over the entire viscosity region from very fluid liquid to the transformation range, with a single value for ΔH_η. On the other hand, as T_0 approaches T_g, the curvature will increase and the difference between ΔH_η for the fluid melt and in the transformation region will become very large.

Fragility of Melts

The large range of variation in the curvature of viscosity/temperature plots has been used as the basis for a system of classification of glassforming melts. Angell has proposed that compositions which exhibit near-Arrhenian behavior over their entire viscosity range be termed

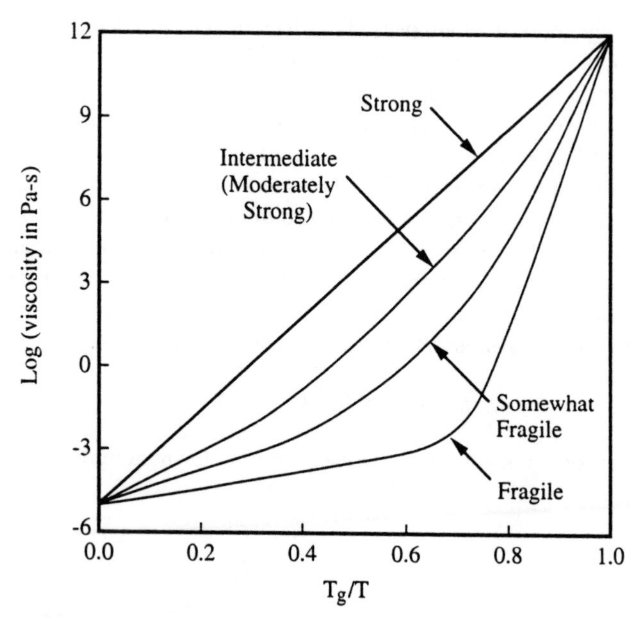

Figure 6.3 *Fragility diagram for typical melts*

strong melts, while those which exhibit a large degree of curvature be termed *fragile melts*. In general, strong melts have well-developed structural units with a high degree of short range order, at least partially covalent bonds, and only gradually dissociate with increasing temperature. Strong melts usually display only small changes in heat capacity upon passing through the glass transformation region. Fragile melts are characterized by less well defined short range order, high configurational degeneracy, and ionic bonds. Their structures disintegrate rapidly with increases in temperature above T_g. Fragile melts are usually characterized by large changes in heat capacity at T_g.

The concept of fragile–strong melt behavior is summarized in a *fragility diagram*, as shown in Figure 6.3. A plot of log viscosity against T_g (which is defined here as the temperature for a viscosity of 10^{12} Pa s) divided by the experimental temperature allows direct comparison of the curvature of various glassforming melts. Examples shown on this diagram range from very strong to quite fragile, with curves for intermediate fragility labeled as moderately strong or somewhat fragile (no terminology has been established for intermediate fragilities). The fragility of a melt can be characterized by consideration of the slope of the viscosity/temperature curve at T_g, *i.e.*, by dividing ΔH_η in the transformation temperature region by T_g in K to obtain the *reduced activation energy for viscous flow*. A *fragility index*, which is equal to the reduced activation energy divided by 2.303 times R, where R is the gas constant, has also been proposed for ranking of melts in order of increasing fragility.

Free Volume Model for Viscous Flow

The temperature dependence of the viscosity can be treated as a function of the free volume of the melt structure. If we write a modified version of the VFT equation, replacing the temperature terms with volume terms, we obtain the expression

$$\eta = \eta_0 \exp[B_1/(V - V_0)] \tag{6.11}$$

where B_1 is a constant, V is the specific volume of the melt, and V_0 is the specific volume for the close-packed melt. If the thermal expansion coefficient of the melt is independent of temperature, this expression is identical to Equation 6.9 since $V - V_0$ is directly proportional to $T - T_0$. This model simply suggests that the viscosity at T_0 is infinite since no molecular motion is possible when the molecules are packed together at their maximum density, *i.e.*, when the specific volume is V_0. Increases in temperature create the free volume, $V - V_0$, which provides space for molecular motion and hence for viscous flow.

Entropy Model for Viscous Flow

A similar equation can be generated from considerations of the effect of temperature on the configurational entropy of a melt. In this approach, the melt is assumed to consist of regions which rearrange as units when they experience a sufficient fluctuation in energy. The sizes of these units are functions of temperature, increasing in size as the configurational entropy decreases and reaching an infinite size when this entropy becomes zero. This model can be mathematically expressed by

$$\eta = \eta_0 \exp(B_2/TS_c) \tag{6.12}$$

where B_2 is another constant and S_c is the configurational entropy. If we replace S_c by the expression

$$S_c = \Delta C_p \left(\frac{T - T_0}{T}\right) \tag{6.13}$$

where ΔC_p is constant with temperature, we obtain an expression identical to Equation 6.9.

COMPOSITIONAL DEPENDENCE OF VISCOSITY

The compositional dependence of the viscosity of glassforming melts is closely related to the connectivity of the structure. In general, changes in composition which reduce connectivity reduce the viscosity, while those which increase connectivity increase the viscosity. These changes are accompanied by changes in fragility which may or may not follow the trend in viscosity, but which are very important in discussion of the temperature dependence of viscosity.

Viscosity data are usually presented in one of two forms. The first form of presentation, which is termed the *isothermal viscosity*, reports the viscosity at specified temperatures. The second form of presentation reports the temperature at which specified viscosities occur, *e.g.*, the values of the Littleton softening temperature or the glass transformation temperature. In general, temperatures referring to a specified viscosity are termed *isokom temperatures* for that viscosity. If a series of curves showing the isokom temperatures are presented on a figure, the individual curves are termed *isokoms* (lines of constant viscosity).

Silicate Melts

Vitreous silica is the most viscous of all common glassforming melts. The glass transformation temperature of vitreous silica, which is strongly influenced by hydroxyl and other impurity concentrations, lies in the range of 1060–1200 °C. The viscosity of silica, which is one of the least fragile melts, varies very slowly with temperature. Production of commercial vitreous silica requires processing temperatures in the range of 2200 °C in order to obtain bubble-free glass.

Addition of alkali oxides to silica results in the formation of nonbridging oxygens and a reduction in the connectivity of the structure. It is not surprising that this reduction in connectivity results in a rapid, monotonic decrease in viscosity with small additions of alkali oxide to silica. The effect of further alkali oxide additions decreases with increasing alkali oxide concentration and eventually becomes quite small for concentrations exceeding 10–20 mol % R_2O. The decrease in viscosity is accompanied by an increase in fragility, which is evidenced by an increase in the reduced activation energy for viscous flow, $\Delta H_\eta / T_g$, in the transformation region and a decrease in ΔH_η in the fluid melt region.

The identity of the alkali oxide present has a relatively small effect on the viscosity of the melt. Although the isothermal viscosity does decrease in the order Cs > Rb > K > Na > Li, the differences among the alkali oxides are small compared to the effect of alkali oxide concentration. In fact, if low viscosity data ($<10^3$ Pa s) are plotted against the concentration of alkali per cm^3 of melt instead of against mol % R_2O, one finds that the data for lithium, sodium, and potassium silicate melts essentially lie on the same line for any given temperature between 1100 and 1400 °C.

Viscosities of melts containing a mixture of two or more alkali oxides are lower than those of corresponding melts containing the same total molar concentration of a single alkali oxide. If we plot isokoms as a function of the ratio of the concentrations of the alkali oxides present for melts containing a constant total concentration of alkali oxides, we might expect a linear variation, or additivity, in the isokom temperature as a function of the composition of the melt. In general, however, we find a small negative deviation from additivity whenever we make such a plot for glasses containing a mixture of alkali oxides. Such a departure from additivity in a property is often considered to be anomalous. Since this particular anomaly is observed in compositions containing a mixture of alkali oxides, it is termed the *mixed alkali effect*.

The viscosities of alkaline earth silicate melts are significantly greater than those of alkali silicate melts. The glass transformation temperature, which occurs at a viscosity around 10^{12} Pa s, is >200 K higher for a calcium silicate glass than for a sodium silicate glass containing the same molar concentration of modifier oxide. Replacement of a modest amount of alkali oxide by an alkaline earth oxide, as is often done in commercial silicate glasses, results in small increases in viscosity due to changes in the field strength of the modifier ion. Although a direct replacement of Na_2O, for example, by an equimolar concentration of CaO does not alter the non-bridging oxygen concentration, the greater field strength of the divalent calcium ion strengthens the bond to neighboring oxygens, thus slightly increasing the strength of the network.

Addition of PbO to silicate melts almost always reduces their viscosity. The viscosity of binary lead silicate melts, for example, decreases monotonically with increasing PbO concentration. These melts also become more fragile with increasing PbO content. Lead oxide also increases the dissolution rate into the melt for impurity particles such as refractory fragments, so that a small concentration of PbO improves the quality of many melts.

Replacement of an alkali or alkaline earth oxide by alumina or gallia reduces the concentration of non-bridging oxygens and increases the connectivity of the network. As a result, the viscosity increases signifi-

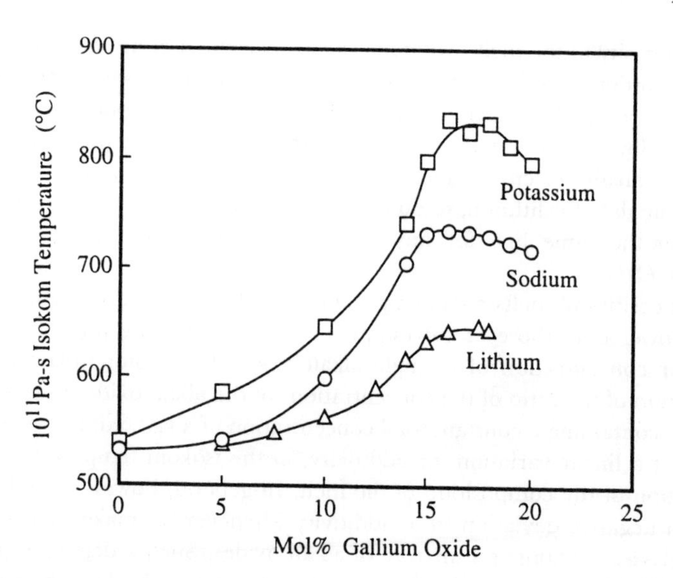

Figure 6.4 *Effect of composition on the 10^{11} Pa s isokom temperature for alkali galliosilicate melts containing 15 mol % of the oxides of potassium, sodium, or lithium*

cantly if these intermediate oxides are added to a melt, as is shown in Figure 6.4. The effect of the intermediate oxide is very small for small concentrations, but then increases rapidly until the concentration of intermediate oxides equals that of the modifier oxides. Further additions of intermediate oxides actually reduce the viscosity slightly as a result of the undetermined structural changes discussed earlier. The presence of intermediate oxides also dramatically increases the difference in viscosity as a function of the identity of the alkali oxide present, as indicated by the divergence of the curves in Figure 6.4.

Viscosity studies have been carried out on a very large number of other silicate systems. The highest viscosities are found for compositions which are free of traditional modifier oxides, *e.g.*, rare earth aluminosilicate melts, and for compositions containing large concentrations of Nb_2O_5 or Ta_2O_5. Low viscosities are usually found for melts containing PbO or a limited concentration of fluorine or other halides.

Borate Melts

The viscosity of vitreous boric oxide is among the lowest reported for common oxide glasses. Boric oxide melts are also considerably more fragile than those of silica, germania, or phosphoric oxide. Since the

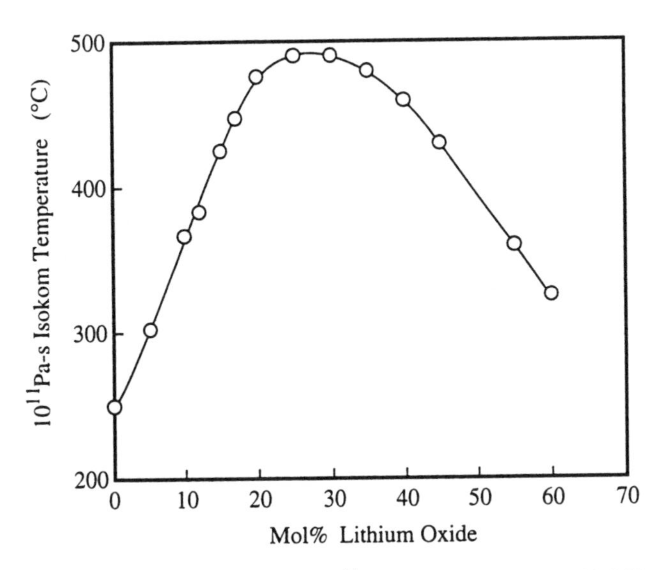

Figure 6.5 *Effect of composition on the 10^{11} Pa s isokom temperature for lithium borate melts*

network of vitreous boric oxide consists of two-dimensional boron–oxygen triangles, with no strong bonding in three-dimensions, the connectivity of the network is low. Many of the boron–oxygen triangles are grouped into boroxol rings. Raman spectroscopy has shown that these boroxol rings dissociate with increasing temperature, resulting in a large thermal expansion coefficient for boric oxide melts. It is probable that the fragility of boric oxide melts is a result of the rapid dissociation of the network which occurs as the boroxol rings open and allow much easier movement within the melt.

Addition of alkali oxides to boric oxide results in considerably more complex behavior than that found for alkali silicate melts. First, even though the connectivity of the melt is increased through conversion of boron–oxygen triangles to tetrahedra with no non-bridging oxygen formation, the fragility of the melt increases with increasing alkali oxide concentration. Second, if we consider the behavior of the viscosity in the transformation region, we find that initial additions of alkali oxide increase the viscosity, while further additions decrease the viscosity, so that maxima in the viscosity *versus* composition curves occur at 25–30 mol % alkali oxide for all five alkali metals (Figure 6.5). If we examine these curves at high temperatures (>1000 °C), where the viscosity is $<10^3$ Pa s, we find that these maxima have disappeared and that the viscosity decreases monotonically with increasing alkali oxide content. Finally, we also find that the viscosity in the transformation region decreases in the order Li > Na > K > Rb > Cs, which is the reverse of the order observed for alkali silicate melts.

Since most common binary borate systems contain large regions of stable immiscibility located near the pure B_2O_3 composition, viscosity data are either not available or are only available in limited compositional ranges for many other binary borate melts. Lead borate melts, for example, are phase separated over the range from just above zero to ≈20 mol % PbO. Transformation region isokoms for melts containing greater concentrations of PbO exhibit maxima at 25–30 mol % PbO which are similar to those observed for the alkali borate melts. These maxima disappear for isokoms for viscosities in the melting range.

The addition of intermediate oxides such as Al_2O_3 and Ga_2O_3 to alkali borate melts has very little effect on the viscosity. The transformation range viscosity still passes through a maximum with increasing alkali oxide content. Similar maxima are also observed for Sb_2O_3–B_2O_3 and Bi_2O_3–B_2O_3 melts for viscosities in the transformation range, even though these binary systems are believed to consist of two network oxides. In fact, the only common binary borate melts which do not display viscosity maxima are those containing either SiO_2 or GeO_2, where the viscosity increases monotonically with increasing concentration of the second glassforming oxide. The viscosity of binary borosilicate melts displays a large, negative departure from additivity, while that of binary borogermanate melts approaches additivity, with only a very small, negative departure from additivity.

One final group of binary borate systems is worthy of note. Glasses can be formed over a wide compositional range in the RF–B_2O_3 systems, where R is any alkali metal ion. The viscosity of melts in these systems exhibits the same behavior as that of the corresponding R_2O–B_2O_3 melts, but the isokom temperatures are considerably reduced for any given viscosity for the same alkali metal ion, as shown in Figure 6.6 for melts containing either Na_2O or NaF. The maxima are less distinct for the alkali fluoroborate melts and may be shifted to lower boric oxide concentrations (comparison of melts containing equal concentrations of the modifier Na^+ ions requires consideration of NaF as if it were actually Na_2F_2). Fluorine acts as a significant flux for these melts, reducing the isokom temperature by over a hundred degrees.

Germanate Melts

Germania melts are quite strong liquids, with viscosities between those of silica and boric oxide. Addition of very small quantities of alkali

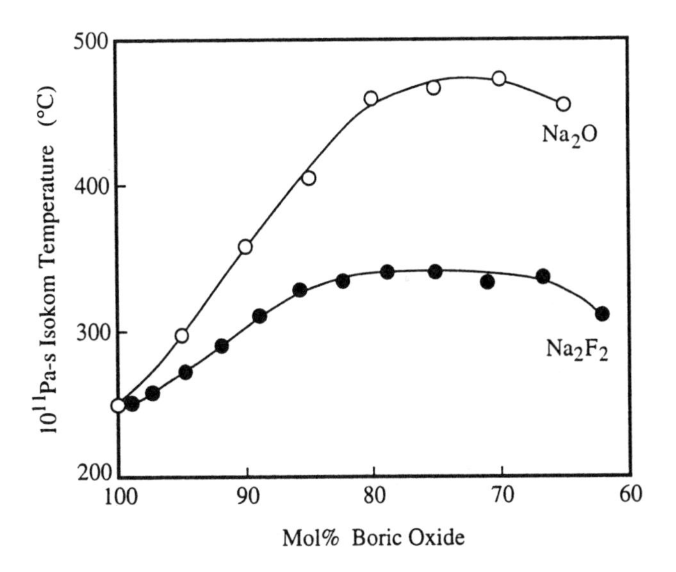

Figure 6.6 *Effect of composition on the 10^{11} Pa s isokom temperature for sodium borate and sodium fluoroborate melts*

oxides to germania results in a sharp decrease in viscosity, with a minimum in viscosity occurring at 1–2 mol % R_2O. Further increases in alkali oxide content initially increase the viscosities of these melts, which pass through maxima at $\approx$ 17 mol % R_2O regardless of the alkali oxide present, and then decrease continually with further increases in alkali oxide concentration (Figure 6.7). The viscosities of these melts are almost independent of the identity of the alkali metal present for compositions containing less than 17 mol % R_2O, but decrease in the order Na > K > Rb > Cs for melts containing more than 20 mol % R_2O. The fragility of these melts increases rapidly with increasing alkali oxide content. Replacement of oxygen by fluorine results in a decrease in viscosity and decreases the sharpness of the maxima at $\approx$ 17 mol % R_2O. Addition of alumina or gallia to potassium germanate melts gradually reduces the size of the maximum, which eventually disappears for melts containing equimolar concentrations of K_2O and Al_2O_3 or Ga_2O_3.

Halide Melts

BeF_2 melts are strong liquids, with relatively low viscosities. Addition of alkali fluorides decreases the viscosity of these melts and increases the fragility of the melt.

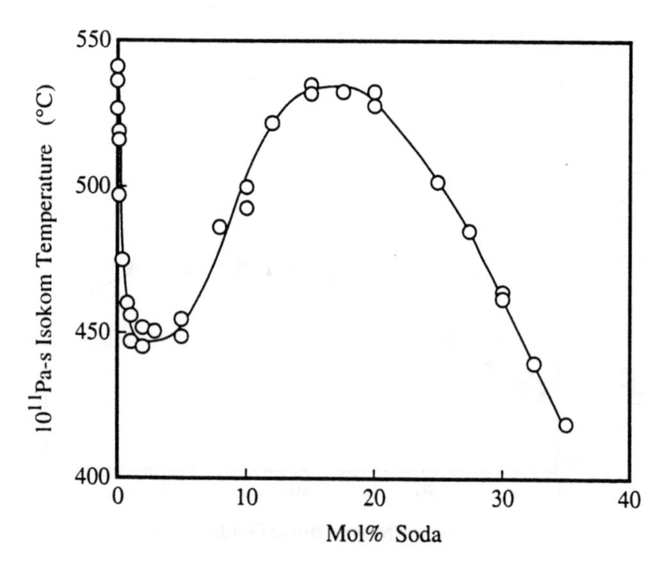

Figure 6.7 *Effect of composition on the* 10^{11} *Pa s isokom temperature for sodium germanate melts*

Heavy metal halide melts are generally very fragile, with very steep viscosity *versus* temperature curves in the transformation range. Differences between glass transformation temperatures and practical melting temperatures are typically only a few hundred degrees. Fluorozirconate glasses, for example, with glass transformation temperatures of ≈300 °C, can be produced by melting at 600–800 °C. Other halides, such as the fluoroaluminates, which have T_g values in the range 400–500 °C, form very fluid melts by 800–1000 °C. Replacement of fluorine by other halides radically reduces the viscosity of these melts, in the order Cl > Br > I. Some halide melts containing iodides can have values of T_g near room temperature.

Chalcogenide Melts

Limited data are available for chalcogenide melts other than elemental S and Se. The viscosity of elemental sulfur, which is very low at its melting point, initially decreases with increasing temperature up to ≈160 °C, where it increases abruptly by several orders of magnitude, reaching a viscosity of nearly 100 Pa s at ≈180 °C. Further increases in temperature result in the expected monotonic decrease in viscosity. This behavior is due to changes in melt structure. The initial melt consists primarily of S_8 molecules in the form of rings. These rings begin to polymerize into long chains containing 10^5–10^6 atoms per chain at 160 °C, leading to a rapid increase in viscosity. Decreases in the average chain length with increasing temperature lead to the decrease in viscosity observed at higher temperatures. This behavior is not observed for selenium, which has a higher viscosity at the crystalline melting temperature and contains shorter chains than those found in molten sulfur at any given temperature.

Effect of Hydroxyl on Melt Viscosities

Hydroxyl serves as a universal flux for glassforming melts. While this contention is based primarily on oxide melts, similar effects have been reported for heavy metal halide melts as well. The magnitude of the reduction in viscosity decreases with increasing temperature for a given composition, with significant effects in the transformation range and much smaller effects in the melting range. It has been shown that the magnitude of the reduction in viscosity is dependent upon the fragility of the melt, with the greatest effect observed for strong melts such as silica and the smallest effect observed for very fragile melts such as sodium metaphosphate and heavy metal fluorides. The change in isokom temperature at 10^{12} Pa s varies from as much as 70 K for silica to only a few degrees for fragile melts. Increases in hydroxyl concentration increase the fragility of the melt.

EFFECT OF THERMAL HISTORY ON VISCOSITY

The viscosity of a melt in the transformation region can be altered by changes in the fictive temperature for homogeneous melts. Since the structural relaxation time in this viscosity regime is significantly long with respect to the experimental time, it is possible to observe changes in the fictive temperature during a viscosity measurement.

If we consider the fictive temperature of a sample to represent a specific structure of the melt, then a different fictive temperature will represent a different structure. Each of these structures will be characterized by its own set of properties. If we alter the surrounding temperature from that of the fictive temperature of the sample, the structure and properties will change to those appropriate for the new temperature (Figure 6.8). The time required for this change will depend upon the viscosity of the melt, which will vary as the fictive temperature changes. The rate of change will initially be determined by the properties of the sample at the initial fictive temperature. Since a higher fictive tempera-

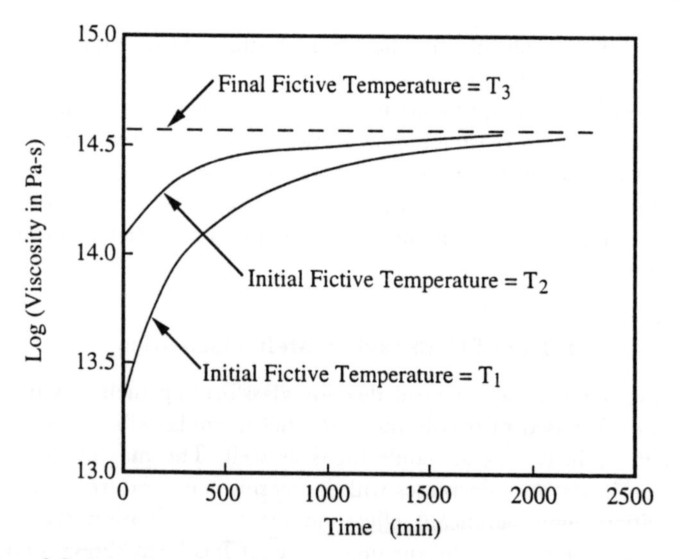

Figure 6.8 *Relaxation of viscosity to a temperature,* T_3, *below the original fictive temperature for two glasses with different fictive temperatures* ($T_1 > T_2 > T_3$)

ture indicates a more open structure and hence a lower viscosity, the initial rate of change will be greater for a sample with a higher fictive temperature. It follows that two samples with fictive temperatures T_1 and T_2, where $T_1 > T_2$, when held at a new temperature, T_3, which is lower than either initial fictive temperature, will relax to the new fictive temperature and eventually have identical properties. Since the sample with an initial fictive temperature of T_1 has a lower initial viscosity (higher fluidity), it will initially relax at a faster rate, but will still require a longer time to achieve equilibrium due to the greater change required to reach this new state.

Suppose we now create two samples of the same melt by holding each at a specified temperature for a sufficient time to reach structural equilibrium at that temperature, *i.e.*, we create two samples with different fictive temperatures. If we now change the temperature of each sample to a new temperature which is intermediate to the two initial fictive temperatures, each sample will approach equilibrium from a different direction (Figure 6.9). The sample with an initial fictive temperature above the new temperature will change faster and reach equilibrium sooner as a result of its initial lower viscosity.

The arguments presented here are acknowledged to be somewhat simplified relative to the changes which actually occur in melts, where a spectrum of relaxation times occur. The behavior of real melts is

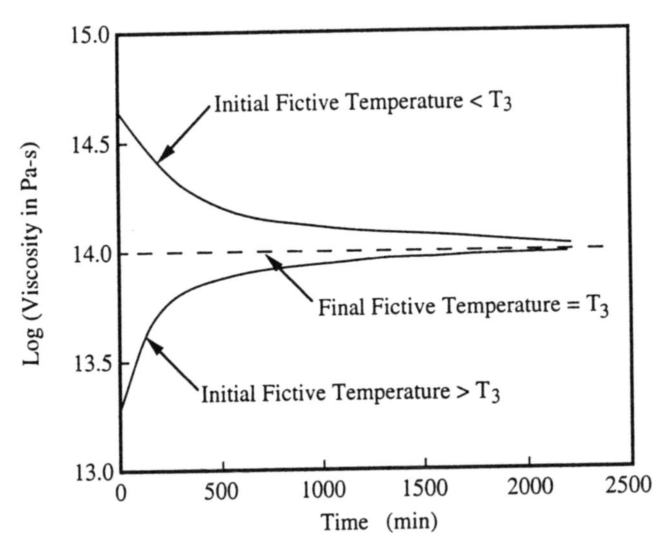

Figure 6.9 *Relaxation of viscosity to a temperature,* T_3, *between the original fictive temperatures of two glasses*

considerably more complex than can be presented within the limited scope of an introductory text. More detailed information regarding relaxation phenomena can be found in the texts by Scherer or Brawer.

EFFECT OF PHASE SEPARATION ON VISCOSITY

Phase separation can radically alter the viscosity of a melt. If stable immiscibility leads to complete separation into two layers of liquid, each layer will be characterized by its own viscosity. Viscosity measurements thus reflect the compositions of the two liquids and have little to do with the bulk composition of the melt. Melts in the calcium borate system have been cooled to room temperature, the two layers of glass separated, and viscosity measurements made on material from each layer. The results, as expected, are independent of bulk composition so long as the composition lies within the stable immiscibility dome and are identical to those obtained for compositions at the ends of the tie line connecting the two phases in equilibrium under the melting conditions.

The effects of phase separation on viscosity are more subtle for compositions exhibiting metastable immiscibility. In many cases, the fine scale of the morphology is such that no indication of phase separation is observed by the naked eye. The measured viscosity will be a function of the compositions of the equilibrium liquids, the connectivity of each phase, and the scale of the morphology. If the phase with the higher viscosity has a connected structure, the viscosity of the lower viscosity will be of little importance since the measured viscosity will be determined by the less mobile phase. On the other hand, if the phase with the higher viscosity exists only as isolated spheres within a matrix of the less viscous phase, the measured viscosity will be near that of the more fluid phase. Any thermal treatment which alters the connectivity of the more viscous phase obviously can radically alter the measured viscosity of the composite material. It follows that viscosity measurements can be used to determine both compositional and temperature limits for the connectivity of the more viscous phase for phase separated melts.

If we heat a sample which has a morphology containing a connected, high-viscosity phase, we can, if the composition lies in the appropriate region of the immiscibility dome, pass the upper limit for connectivity of the high-temperature phase. If this occurs, the connectivity of that phase will rapidly decrease and the measured viscosity will revert to that of the lower viscosity phase. Since that viscosity is typically several orders of magnitude below that of the more viscous phase which had been controlling the measured viscosity, the viscosity of the sample will appear to abruptly decrease and the melt will suddenly become much more fluid. This effect can occasionally have unfortunate side effects if the sample now flows much more rapidly than anticipated.

The relation between phase separation and viscosity has been studied as a function of time during which the scale of phase separation increases. Under these circumstances, the viscosity has been observed to increase by as much as five orders of magnitude during an isothermal heat treatment for times of several hundred hours. The viscosity initially changes rapidly as the connectivity of the structure and the compositions of the equilibrium phases approach their final values, and then more slowly as coarsening, or growth in the scale of the microstructure, occurs.

EFFECT OF CRYSTALLIZATION ON VISCOSITY

Examination of the literature reveals that most viscosity data are reported in either the glass transformation (10^{12}–10^{8} Pa s) or the melting ($<10^{5}$ Pa s) range. Relatively little data can be found for viscosities between these regions. The paucity of data within the softening region of the viscosity *versus* temperature curve is primarily due to the tendency for melts to crystallize within this viscosity range. Data can normally be obtained only for compositions which are significantly more stable against crystallization than most melts.

The effect of crystallization on the measured viscosity, which is not equal to the true viscosity of the melt, depends upon the details of the crystallization process. If a sample crystallizes *via* growth of crystals from the surface, the sample is essentially encapsulated in a shell of crystals which prevent deformation and the viscosity appears to increase to infinity (or at least to some value orders of magnitude greater than the viscosity of the encapsulated melt). Crystals distributed throughout the bulk, however, do not necessarily prevent deformation, so that the sample usually still exhibits signs of viscous flow. If the crystals have the same stoichiometry as the melt, very little effect will be noticed until the crystals begin to touch and hence to interfere with flow. If the crystals have a different stoichiometry from that of the melt (the most common case), it is possible for the true melt viscosity either to increase or to decrease as crystallization occurs, depending upon the relative compositions of the melt and the crystals. The initial effect on the measured viscosity is thus somewhat difficult to predict. Once the crystals impinge, however, the apparent viscosity will increase due to the blocking effect of the crystals. Eventually, flow will cease as the sample becomes fully crystalline.

The shape of the crystals has some effect on the behavior considered here. Spherical crystals will have less effect on deformation than plate-like or needle-like crystals. The minimum volume fraction of crystals which significantly affect the viscosity will thus be a function of the shape and size of the crystals as well as their composition and location within the sample. Since very little information regarding these effects has been published, details of the effects of crystallization on viscous deformation of melts remains largely a mystery.

SUMMARY

The viscosity of a glassforming melt varies over 14–15 orders of magnitude during the production of a glass. As a result, several different techniques must be used to measure a complete viscosity/temperature curve. The viscosity/temperature curve has a complex shape, which is commonly described by the Vogel–Fulcher–Tamman equation or by discussion of the fragility of the melt. Compositional changes which decrease the connectivity of the vitreous network decrease the viscosity, while changes which increase the connectivity increase the viscosity. Hydroxyl, fluorine, and PbO all act as fluxes and reduce the viscosity of oxide melts. The viscosity in the glass transformation range is a strong function of the fictive temperature of the melt and changes with time if a specimen is heat treated at a temperature different from its original fictive temperature. Phase separation and crystallization strongly alter the viscosity of melts, especially in the glass transformation range. The connectivity of the vitreous phases is especially important in determining the effective viscosity of phase separated specimens.

Chapter 7

Density and Thermal Expansion

INTRODUCTION

Trends in the density, thermal expansion coefficient, refractive index, and viscosity of glasses as a function of bulk composition serve as the basis for many of the common structural models used today. These models were generated long before Raman, NMR, and other modern spectral techniques were developed. While details of these models have been refined using more sophisticated methods, the basic concepts of network structures, bridging and non-bridging oxygen formation, and changes in coordination number with changes in composition were originally proposed in an attempt to explain trends in property behavior. This approach to glass structure remains common even today, with many structural models proposed on the basis of property studies and later confirmed by the results of spectral studies.

TERMINOLOGY

The *density* of a material is defined as the mass of the substance per unit of volume, or

$$\rho = \frac{M}{V} \tag{7.1}$$

where ρ is the density, M is the mass, and V is the volume of the sample. If the sample is free of bubbles, voids, or other defects, the calculated density is the *true density* of the material. If, however, the sample contains bubbles, which is occasionally the case for glasses, the calculated density will be less than that of the true density and is termed the *apparent density*. Inclusions with higher densities than the true density, which might, for example, be due to particles of unmelted batch or crystals formed during cooling, will cause the apparent density to be greater than the true density.

The *molar volume* is defined as the volume occupied by one mole of a material and is obtained by dividing the molecular weight of a material by its density, or

$$V_m = \frac{MW}{\rho} \tag{7.2}$$

where V_m is the molar volume, MW is the molecular weight of the substance, and ρ is the true density of the material. Since the density of a material is sensitive to both the volume occupied by the atoms and to their mass (atomic weight), molar volume is often used to compare the behavior of glasses. In many cases, seemingly anomalous behavior in density is readily explained by consideration of the molar volume.

The method used to represent the formula of a glass can alter the calculated V_m by altering the value used for the molecular weight. The most commonly used definition of molecular weight is based on the oxide formulation method with mole fractional contributions from each component. The molecular weight of a $25Na_2O$–$75SiO_2$ glass is equal to the sum of 0.25 times the molecular weight of Na_2O plus 0.75 times the molecular weight of SiO_2. If we use molecular weights of 61.981 for Na_2O and 60.084 for silica, we obtain a molecular weight of 60.558 for this glass. Using a measured density of 2.434 g cm^{-3}, we find that the molar volume of this glass is 24.88 cm^3 mol^{-1}. While molar volume is rarely calculated on any basis other than the mole fractional formula used here, it should be recognized that other methods were used in the past and are sometimes still used today.

The *thermal expansion coefficient* of a material is a measure of the rate of change in volume, and therefore density, with temperature. The *true* (sometimes called *instantaneous*) *thermal expansion coefficient* is defined as the slope of the volume *versus* temperature curve at a specified temperature and constant pressure (usually 1 atmosphere), or

$$\alpha_v = \frac{1}{V}\left(\frac{\partial V}{\partial T}\right)_P \tag{7.3}$$

where α_v is the true volume expansion coefficient, V is the volume of the sample, and $(\partial V/\partial T)_P$ is the slope of the curve. The *average*, or mean, *thermal expansion coefficient*, $\bar{\alpha}_v$, which is much more commonly reported, is defined by the change in volume, ΔV, over a specified temperature interval, ΔT, or

$$\bar{\alpha}_v = \frac{1}{V}\left(\frac{\Delta V}{\Delta T}\right) \tag{7.4}$$

Although the thermal expansion coefficient is actually defined in terms of the volume of the substance, this value is somewhat difficult to measure. As a result, the expansion coefficient for glasses is usually only determined in one direction, *i.e.*, the measured value is the *linear thermal expansion coefficient*, α_L. The true and average linear thermal expansion coefficients are given by Equations 7.3 and 7.4, respectively, where V is replaced by L in each equation. Since glasses are usually isotropic materials with relatively small thermal expansion coefficients, $\alpha_v = 3\alpha_L$ can be used to approximate α_v with very little error in calculation.

Virtually all reported thermal expansion coefficients for glasses are actually average linear thermal expansion coefficients over some specified temperature range. The particular temperature range represented by this value is not always specified in the reported data. Data for commercial glasses are usually obtained for either the range from 0 to 300 °C, 20–300 °C, or 25–300 °C. Data for experimental studies may be reported for almost any temperature range, so caution must be used when comparing results from different studies. Since the true thermal expansion coefficient can be a strong function of temperature, knowledge of the temperature range used to define an average thermal expansion coefficient is vital for application of the data.

Since most linear thermal expansion coefficients lie between 1 and 50 $\times\ 10^{-6}$/K, metallurgists, ceramists, and other material scientists usually report values with units of ppm K^{-1}. Traditionally, however, glass technologists used $10^{-7}\ K^{-1}$ as the basis for reporting thermal expansion coefficients. A glass technologist might, therefore, indicate that the linear thermal expansion coefficient for a certain glass is 86, while a ceramist might indicate the same coefficient as 8.6. Since older data were frequently reported in terms of °F instead of K, considerable care must be taken when using published thermal expansion coefficients to insure that the values used are actually those intended in the original source.

MEASUREMENT TECHNIQUES

Density

The most direct method for determining density involves weighing a sample of known geometry, calculating its volume from its dimensions, and using Equation 7.1 to calculate the density. If the available samples do not have simple geometries, we can use Archimedes' principle to determine the volume by liquid displacement. The sample is weighed both in air and suspended in a liquid of known density. The difference in weight equals the weight of the displaced liquid. Since we know the density of the liquid, ρ_L, we can calculate the displaced volume using Equation 7.1. Dividing the weight of the sample in air, W, by the volume of liquid displaced then yields the density of the sample. The density is calculated from the expression

$$\rho = \frac{W\rho_L}{(W - W_s)} \tag{7.5}$$

where W_s is the suspended weight of the sample. The choice of the immersion liquid is based on convenience and the chemical durability of the sample. Water is usually used where possible, but kerosene or an alcohol are often used for samples which react with water.

Densities of glasses are frequently measured by suspension in heavy liquids. A density gradient column is formed by introducing two liquids of different densities in continually varying proportions into a tall glass cylinder. A set of standards of known densities are then placed into the column. Since each standard will float at a depth where the liquid density equals that of the standard, a calibration graph of density *versus* depth can be plotted. A sample is placed in the liquid and allowed to come to rest at an equilibrium depth. The density is then read from the calibration plot. If the densities of the two liquids are very similar, these measurements are capable of detecting density differences of as little as 1 ppm among a set of nearly identical samples.

Sink–float methods are commonly used for quality control measurements for glasses. A sample is placed into a test tube containing an organic liquid which is slightly more dense than the sample. The tube is heated until the density of the liquid becomes less than that of the sample, whereupon the sample will begin to sink. If the test tube also contains a standard of known density, the difference in temperature at which the sample and standard sink can be used to calculate the difference in their densities, provided the temperature dependence of the density of the liquid is known. This method is capable of detecting differences in density of as little as 20 ppm.

Thermal Expansion Coefficients

Almost all reported thermal expansion coefficients for glasses have been obtained using some variation of a push-rod dilatometer. In its simplest form, a push-rod dilatometer consists of a cylinder of a material of known thermal expansion coefficient which is fixed in place at one end and surrounded by a heating device. A sample is placed inside and against the end of this cylinder. A rod of the same material as the cylinder is placed against the sample. The other end of the rod is connected to some device capable of measuring very small changes in the position of the end of the rod. Heating the region containing the sample results in expansion of the surrounding cylinder, the rod, and the sample. If the sample has a different thermal expansion coefficient from that of the apparatus, the end of the rod will be displaced by an amount determined by the sample length and by the difference in thermal expansion coefficients between the sample and the apparatus material. Determination of the true thermal expansion coefficient of the sample requires correcting the displacement *versus* temperature data for the expansion of the apparatus.

Most dilatometers used in studies of glasses are constructed from vitreous silica. Since the average linear thermal expansion coefficient of vitreous silica is only about 0.55 ppm K^{-1} over typical temperature ranges covered in these measurements, the correction factor for the apparatus expansion is quite small. Furthermore, since virtually all glasses have glass transformation temperatures less than that of vitreous silica, the upper use temperature of $\approx 1000\,°C$ imposed by the viscosity of this material is rarely of concern.

A more sensitive and accurate dilatometer can be constructed by use of two push-rods in an arrangement which measures the difference between the expansions of the sample and a reference standard. This design is particularly useful for studies of glasses to be used for sealing to another material, since the two materials to be joined can be used in the sample and reference positions and the difference between their expansions can be determined directly.

The most accurate thermal expansion measurements are obtained using interferometry. These instruments are capable of detecting changes in dimension of as little as 20 nm using visible light. While not used widely for routine measurements, NIST (National Institute of Standards and Technology) expansion standards are calibrated using these techniques. Readers interested in these methods should consult a thermal analysis text for details.

DENSITY AND MOLAR VOLUME

The density of a glass is a strong function of its composition. Density is also dependent to a lesser degree on the measurement temperature and the thermal history of the sample. Changes in morphology can have a small effect on density for phase separated glasses. Crystallization of a glass can significantly alter the density if the density of the crystalline phase is very different from that of the residual glass.

Compositional Effects

Densities of the common glassforming oxides are less than those of the corresponding crystalline forms of these compounds. Vitreous silica, for example, has a density of 2.20 g cm^{-3}. This value can be compared with the densities of α-quartz (the room temperature crystalline form of silica) of 2.65 g cm^{-3}, β-cristobalite (the least dense crystalline form of silica) of 2.27 g cm^{-3}, and coesite (a very dense crystalline form of silica obtained at high pressures) of $\approx$3.0 g cm^{-3}. If we calculate the *free volume*, V_f, of the glass using the simple relationship

$$V_f = 1 - \frac{V_x}{V_g} \tag{7.6}$$

where V_x is the molar volume of the crystalline form and V_g is the molar volume of the glass, we obtain a value of 0.27, or 27%, for vitreous silica if we base our calculation on the dense crystal coesite. This large free volume implies that the glass has a very large fraction of interstitial space within the network for accommodation of other ions such as the monovalent alkali ions and the divalent alkaline earth ions.

If the networks formed by the primary glassforming oxides contain a large number of empty interstices, it should be possible to stuff a correspondingly large number of modifier ions into these interstices. Such a process would increase the mass of a substance without increasing its bulk volume, resulting in an increase in density. (Imagine the change in a dry sponge after soaking in water. Assume that the sponge does not swell as it absorbs the liquid.) Indeed, we find that the addition of alkali ions to any of the common glassforming oxides results in an increase in density (Figures 7.1 and 7.2). Even Li_2O, which has only half the molecular weight of silica, increases the density of silicate, borate, or germanate glasses when substituted for the basic glassforming oxide.

This simple picture for the density of glasses has several flaws. Carefully examine Figures 7.1 and 7.2. Note the order of increase of density with increase in the atomic weight of the alkali ion present. The glasses containing lithium are often more dense than glasses containing sodium or potassium. In some cases, glasses containing sodium are more dense than those containing potassium. These trends are not consistent

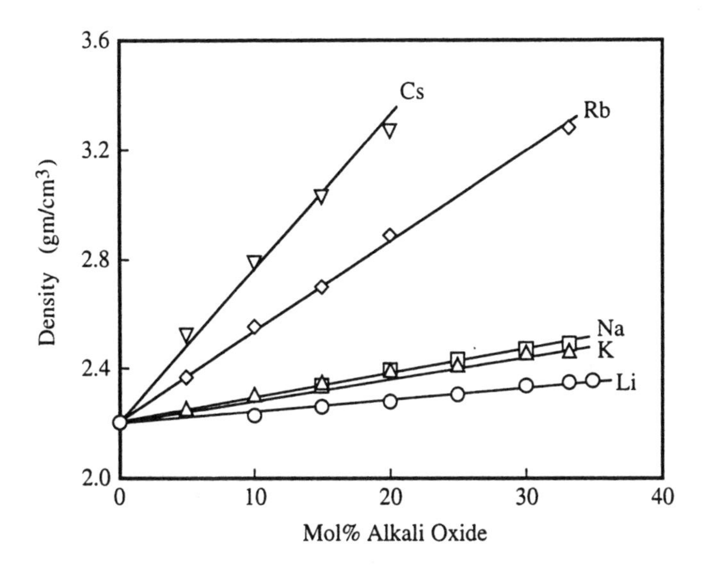

Figure 7.1 *Effect of composition on the density of alkali silicate glasses*

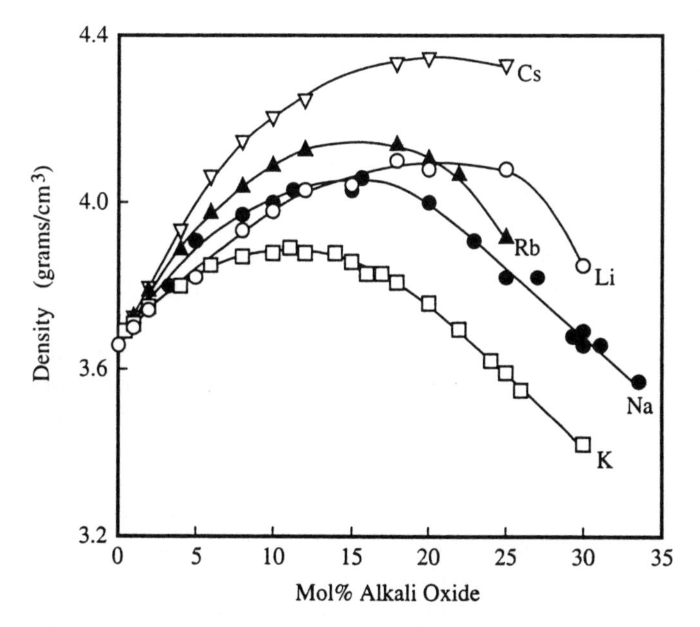

Figure 7.2 *Effect of composition on the density of alkali germanate glasses*

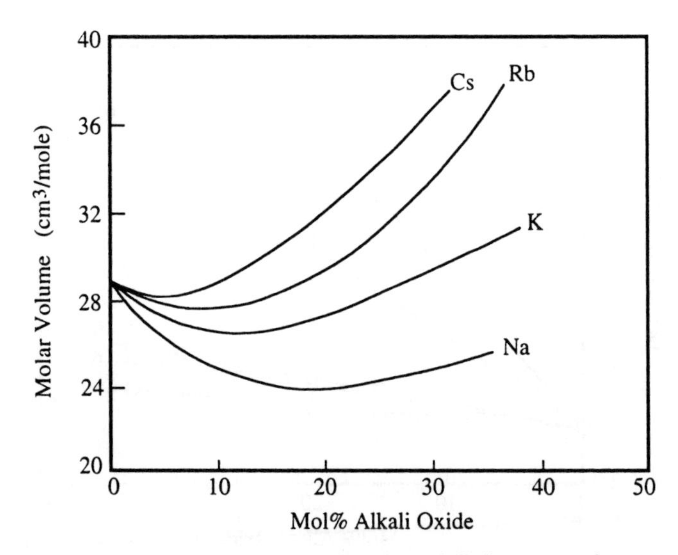

Figure 7.3 *Effect of composition on the molar volume of alkali germanate glasses*

with the simple interstice-filling model discussed above. If a potassium ion weighs about six times as much as a lithium ion, and both simply occupy interstices in an existing network, how can the lithium glass be more dense?

If we use the density data in these figures to generate molar volume data, the unusual nature of the trends in density disappear. The molar volumes in every case increase in the order Li < Na < K < Rb < Cs (as example is shown in Figure 7.3 for the alkali germanate glasses). Additions of lithium or sodium to the network reduce the molar volume, implying that they cause a shrinkage of the network. Potassium, rubidium, and cesium, on the other hand, increase the molar volume of glasses, implying that they force an increase in the volume of the structure.

Let us consider what actually happens when a glass forms from a melt. At very high temperatures, where the melt is fluid, the melt consists of a mixture of modifier ions and various structural units of the network, which may be as small as the basic building block of the network or which may consist of a few of these building blocks connected to form small discrete ions. As we cool the melt, the network begins to form as the structural units become connected. If no alkali or other modifier ions were present, the network could form without hindrance from these ions. When modifier ions are present, however, the network must form around these ions. If the modifier ions are larger than the interstices which would form in their absence, the network will now contain a number of interstices which are larger than those of the modifier-free structure. If the modifier ions are smaller, their attraction to the oxygen ions can lead to a decrease in the size of the interstices. In other words, the modifier ions are not simply filler for predetermined interstices, but modify the interstices actually formed. The relatively high mobility of the modifier ions on the time scale of network relaxation as the melt approaches the transformation range actually allows each modifier ion to influence the size of several interstices. As a result, large ions such as cesium may increase the free volume of the network, while small, higher field strength ions such as lithium can decrease the free volume.

The densities and molar volumes of glasses in binary systems containing two glassforming oxides (B_2O_3–SiO_2, B_2O_3–GeO_2, GeO_2–SiO_2) vary almost linearly with the mol % of each component in the glass. While there is some evidence for small negative departures from additivity in the glasses containing boric oxide, the difficulty in defining the correct density of vitreous boric oxide prevents placing very much emphasis on such small deviations. (Vitreous boric oxide is particularly susceptible to the effects of thermal history on density which will be discussed later in this chapter.) These trends suggest that the packing of the basic building blocks in these structures is almost unaffected by the availability of two units of different sizes and dimensionality.

The most dense oxide glasses contain very high concentrations of PbO or Bi_2O_3, with maximum values in the range of 8.0 g cm^{-3} for glasses in the PbO–Bi_2O_3–Ga_2O_3 system. The rather low maximum densities obtainable for oxide glasses are due to the averaging effect of the low atomic weight oxygen ions, which constitute at least 50 atomic % of the glass.

Heavy metal fluoride glasses, which contain a number of high atomic weight cations, are typically less dense (4–6 g cm^{-3}) than heavy metal oxide glasses due to the averaging effect of the higher atomic % of fluorine ions in their structures. Increasing the atomic weight of the halide ion used in the glass, *e.g.*, replacing fluorine by chlorine, might be expected to increase the density of these glasses. In many cases, however, the replacement of the smaller fluorine ions by larger chlorine ions results in an increase in molar volume which offsets the increase in atomic weight and actually causes a decrease in density of the glass.

Chalcogenide glasses typically have densities of the order of 3–5 g cm^{-3}, although glasses containing thallium can have densities greater than 6 g cm^{-3}. The density of vitreous selenium is 4.29 g cm^{-3}. Addition of arsenic or germanium to selenium results in very small increases in density, while addition of phosphorus results in a large decrease (a factor of almost 2) in density. As might be expected from their atomic weights, replacement of selenium by sulfur decreases the density of arsenic selenide glasses, while replacement of selenium by tellurium increases their density. The molar volumes of a large range of chalcogenide glasses containing Ge, Sb, Se, As, and Te all lie in the range of 17–20 $cm^3 mol^{-1}$.

Thermal History Effects

While the densities of crystalline materials are not particularly sensitive to the thermal history of the sample, densities of glasses are always dependent upon the thermal history of the particular sample measured. Although the differences in density which result from changes in thermal history are not particularly large, they can be very important in certain applications, especially those requiring highly reproducible values of the refractive index of glasses.

Changes in density due to changes in thermal history are best explained by consideration of the volume/temperature diagram shown in Figure 7.4. A melt contracts as it is cooled. Once the temperature of the melt enters the transformation region, the relaxation times becomes significant compared to the rate of cooling. If the melt is cooled rapidly, complete structural rearrangement to the state appropriate for each temperature cannot occur during the time the melt is actually at that temperature. Once the structural relaxation time exceeds the characteristic time for the experimental cooling rate, the structure becomes effectively fixed and does not change any further with decreasing temperature. The temperature which characterizes the structure of the equilibrium liquid which is frozen into the glass is known as the *fictive temperature*. Cooling of the melt at a slower rate will allow equilibrium to be maintained to a lower temperature, where the structure is more dense, before the structure become fixed. The fictive temperature will be lower and the glass, after cooling through the elastic contraction region to room temperature, will be more dense.

Since the density of a glass is controlled by the cooling rate, a plot of density *versus* cooling rate can be generated. The density becomes independent of the cooling rate for very fast rates. If the furnace temperature is decreased very rapidly, the low thermal conductivity of the melt will prevent the sample from maintaining equilibrium with its surroundings. In this case, the thermal inertia of the sample will effectively determine the fastest rate of cooling which actually occurs,

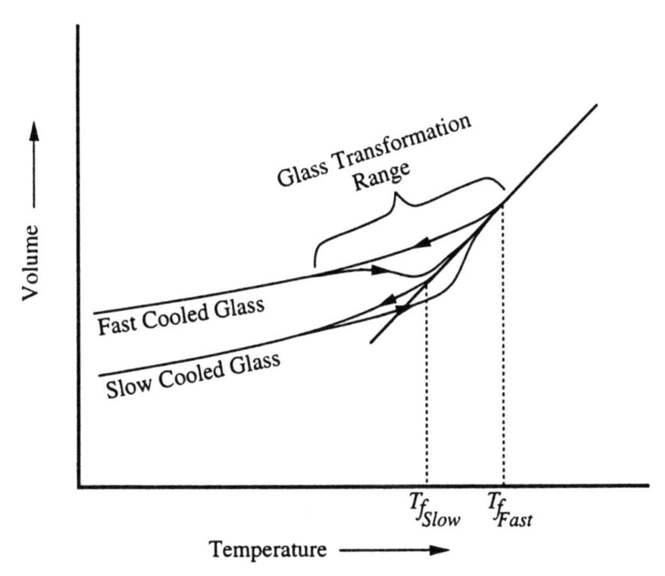

Figure 7.4 *Effect of temperature on the volume of glassforming melts*

regardless of the rate of cooling of the surroundings of the sample. This maximum cooling rate will then determine the lowest density possible for a given sample and will depend upon the size and shape of the sample.

Density can also be altered by heat treatment at a specific temperature for times sufficient to allow equilibration of the structure to that appropriate for the treatment temperature. If we rapidly quench a melt, we obtain a glass sample which has a high fictive temperature and a low density. If we reheat this sample to a temperature within the transformation range, but below the original fictive temperature, the sample will readjust to the structure appropriate for the new temperature, become more dense, and will now be said to have a new fictive temperature. We can also slowly cool a melt, producing a low fictive temperature, and then reheat the sample to a higher temperature. In this case, the sample will expand and readjust to the new, higher fictive temperature. If two samples are cooled at very different rates and then reheated to an intermediate temperature, the densities will change in opposite directions as they readjust to the new fictive temperature, as is shown in Figure 7.5, which resembles the corresponding figure for changes in viscosity discussed in Chapter 6.

Although most glasses behave in the manner discussed here, a few exhibit anomalous density changes with fictive temperature. While the density of most liquids decreases monotonically with increasing tempera-

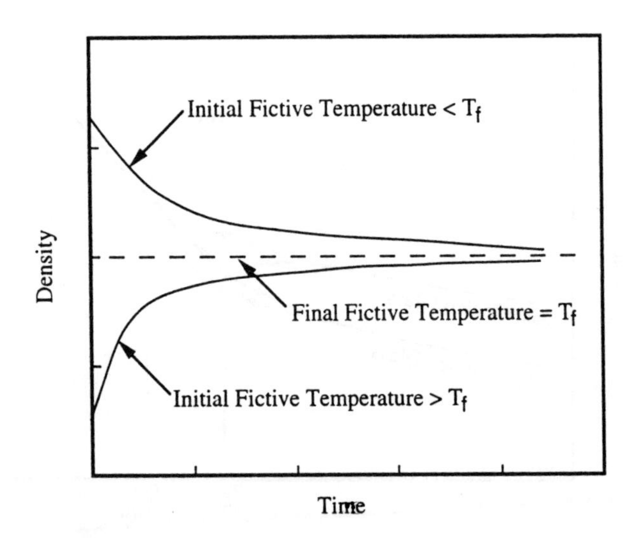

Figure 7.5 *Effect of time on the density of glasses for temperatures in the glass transformation region*

ture, the densities of a few liquids, *e.g.*, water, and a few glassforming melts, *e.g.*, pure silica, actually pass through a maximum with increasing temperature. The unusual temperature dependence of the volume of the equilibrium liquid for silica in the region from 1000 to 1500 °C produces the unusual curve shown in Figure 7.6 for the room temperature density of vitreous silica as a function of fictive temperature. The addition of even small amounts of other oxides to silica eliminates the maximum in density (minimum in volume) of the liquid and causes the density to behave in the usual manner.

Effect of Phase Separation and Crystallization

Phase separation has very little effect on the bulk density of glasses. Since the measured density is a volume average of the densities of the phases present, the apparent density will vary smoothly as the relative volumes of the two phases vary. Plots of density *versus* composition expressed as weight fraction of one component for binary glassforming systems will always display a slight negative curvature within a region of phase separation. While such a region must occur for phase separated glasses, the existence of such a region is not proof that phase separation exists. The magnitude of the negative curvature is often so

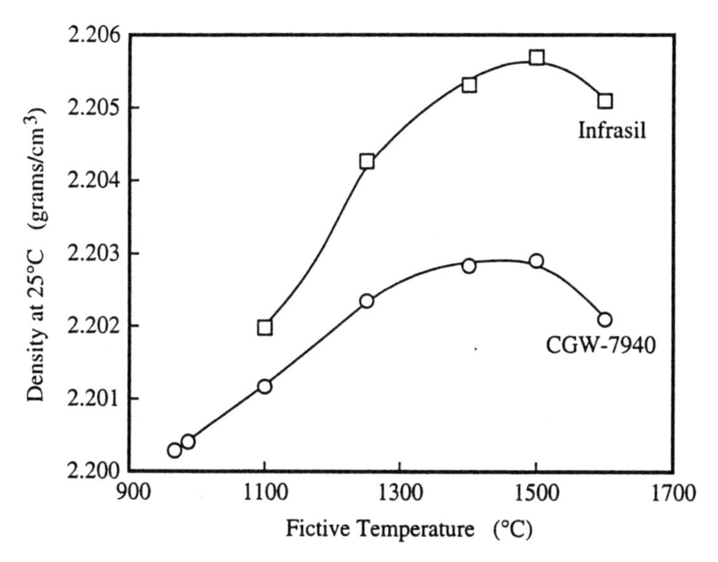

Figure 7.6 *Effect of fictive temperature on the density of vitreous silica*

small that density plots are of little value in determining potential phase separation regions.

Crystallization can lead to much larger changes in density of glasses. If the density of the crystalline phase is significantly different from that of the residual glass remaining after crystallization, the density of the composite can vary over a large range. The density of commercial lithium aluminosilicate glass-ceramics, for example, is about 5% greater than the density of the glass before crystallization. The densities of glass-ceramics produced from lithium silicate glasses are not only greater than those of the glasses, but also differ with the identity of the crystalline phase formed, *i.e.*, lithium metasilicate or lithium disilicate. If the density of the crystalline phase is less than that of the glass, the density of the glass-ceramic may decrease when the sample in crystallized.

As a first approximation, the density, ρ, of a glass-ceramic is given by the expression

$$\rho = \rho_x V_x + \rho_g V_g \tag{7.7}$$

where ρ_x and ρ_g are the densities of the crystalline and glassy phases, respectively, and V_x and V_g are the volume fractions of the same phases. This expression is usually correct to within $\approx 1\%$ measured density. Deviations from this expression are often due to microcracks or to small concentrations of a third phase which is not included in the calculation.

Radiation Effects

Exposure of silicate glasses to high energy radiation usually results in compaction of the glass, with density increases of the order of 1% for very high radiation doses. This compaction occurs at room temperature and is quite stable over long times, with no evidence of room temperature annealing. The density reverts to the pre-irradiation value, however, when the sample is heated, even though the heat treatment temperature may be $<T_g$.

Most studies have dealt exclusively with vitreous silica, which compacts by only a few hundred ppm at doses as great as 10^{10} rad. Some commercial borosilicate glasses compact up to 40 times as much as vitreous silica for comparable radiation doses. Limited data for alkaline earth aluminosilicate glasses suggests that they expand slightly under these same conditions.

The radiation-induced compaction of vitreous silica can be reversed by saturating the sample with molecular hydrogen prior to irradiation. After irradiation, the sample is found to expand by a few hundred ppm (doses of 10^{10} rad). Large quantities of Si–OH and Si–H bonds are formed in the irradiated glasses. Apparently the production of the broken bonds due to formation of bound hydrogen species allows the structure to relax and expand, reducing the density.

Pressure Compaction

Glasses can also be permanently compacted by application of very high pressures. The density of vitreous silica can be increased by up to 15% by pressures in the 100 kbar range at room temperature or by lower pressures at elevated temperatures. Densification is facilitated by a high shear component in the applied stress and by heating to within the glass transformation range. Annealing of these density increases occurs at temperatures well below the transformation range, but the kinetics of the annealing process are very complex.

THERMAL EXPANSION BEHAVIOR

The thermal expansion curve for a glass yields three important pieces of information: the thermal expansion coefficient, the glass transformation temperature, and the dilatometric softening temperature. The thermal expansion coefficient indicates the relation between the volume of a glass and its temperature. The glass transformation temperature indicates the onset of viscoelastic behavior, while the dilatometric softening temperature indicates the onset of flow under a modest load. Each of these properties is a strong function of glass composition. Lesser effects are due to changes in thermal history or the heating rate used during the measurement. The morphology of a sample has, at best, only a very small effect on the thermal expansion coefficient for phase separated glasses, while the glass transformation and dilatometric softening temperatures are strongly affected by phase separation.

Crystallization of a glass can also significantly alter the thermal expansion behavior of a glass.

Fundamentals of Thermal Expansion Behavior

Thermal expansion of close packed structures such as NaCl is the direct result of an increase in bond length with increasing temperature. The increase in bond length arises from the asymmetry of the potential energy *versus* interatomic distance curve of the Condon–Morse potential energy diagram (Figure 7.7). This curve results from the interaction of the repulsive and attraction terms of the interatomic potential, E_v, which is often expressed as

$$E_v = -\frac{A}{R^n} + \frac{B}{R^m} \tag{7.8}$$

where A, B, n, and m are constants and R is the interatomic distance.

If the potential energy well were symmetric, the average interatomic distance would be independent of temperature even though the magni-

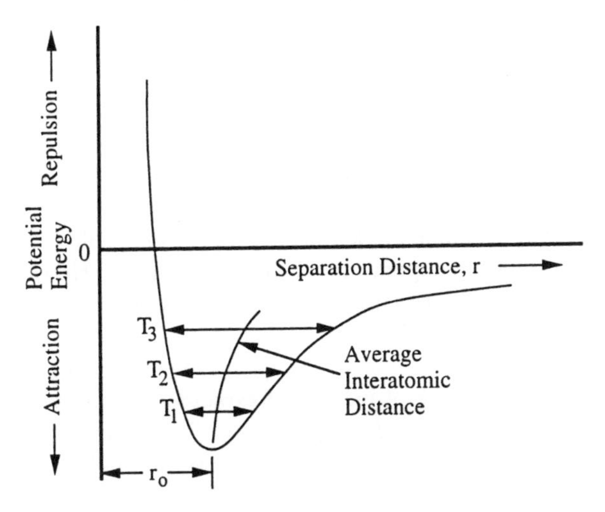

Figure 7.7 *Condon–Morse cuve illustrating the cause of the thermal expansion of bonds*

tude of the vibration amplitude would increase. Since the curve is asymmetric, however, the average interatomic spacing increases with increasing temperature. The degree of asymmetry is dependent upon the value of n, which increases with bond strength from metallic to ionic to covalent bonds. As n increases, the curve becomes more symmetric and the thermal expansion coefficient decreases. Other factors also affect the symmetry of the bond. The creation of a non-bridging oxygen from a bridging oxygen increases the asymmetry of the bond to the neighboring silicon or other network cation. Bonds between an anion and two neighboring cations of different field strength are less symmetric than bonds involving neighboring cations of equal field strength.

While the simple lattice vibration model works well for close packed structures, additional processes can occur in less tightly packed network structures such as those found in glasses. Bond bending can alter the positions of atoms, as can rotation about an axis. These processes can counteract the expansion of the bond length due to the increased amplitude of vibration, resulting in very low expansion coefficients. Filling of interstices inhibits these processes and tends to cause an increase in thermal expansion coefficients.

Eventually, a temperature is reached where bond breaking begins to add still another factor to the overall thermal expansion of a structure. Bond breaking occurs at the melting point for crystalline materials, where a large, discontinuous increase in the thermal expansion coefficient usually occurs. The corresponding effect begins to occur at a measurable rate as we enter the glass transformation range during heating of a glass, where we find a gradual increase in the thermal expansion coefficient to the value of the equilibrium liquid. The intercept temperature of the extrapolated slopes from the expansion curve in the elastic and melt regions is usually used as the glass transformation temperature for thermal expansion data (Figure 7.8). As discussed earlier, the glass transformation temperature usually occurs when the viscosity is around 10^{12} Pa s.

If the melt were contained in a crucible, expansion would continue to occur to the highest temperature of the experiment. However, the sample used in thermal expansion measurements is actually under a small load introduced by the push rod. Eventually, deformation will occur as the push rod sinks into the sample or the sample bends under the load. The temperature of maximum expansion is called the *dilatometric softening temperature*, or T_d (Figure 7.8). This temperature is an artifact of the dilatometric method used to measure the thermal expansion coefficient. As a result, it is slightly dependent upon the load applied by the push rod and the cross-sectional area of the sample, which, in

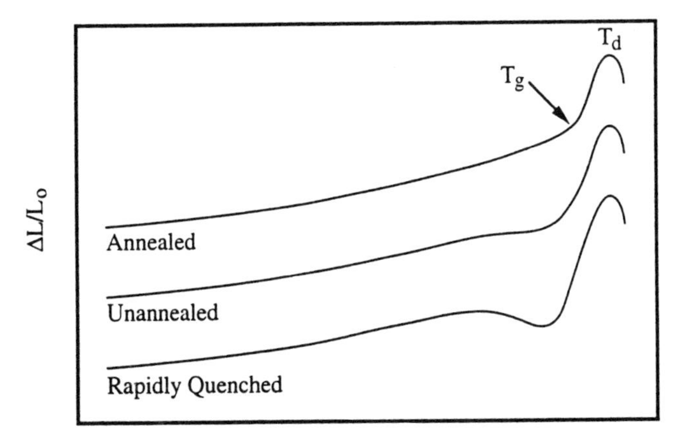

Figure 7.8 *Thermal expansion curves for annealed and quenched glasses*

combination, determine the stress on the sample. Values obtained from most laboratories, surprisingly, agree within a few degrees.

Compositional Effects on Thermal Expansion Coefficients for Homogeneous Glasses

Thermal expansion coefficients of glasses, with few exceptions, increase with increasing temperature. The data can often be expressed over a wide temperature range for the elastic material by an equation of the form

$$\alpha = \alpha_0 + \alpha_1 T + \alpha_2 T^2 + \alpha_3 T^3 \tag{7.9}$$

where α_0, α_1, α_2, and α_3 are experimental constants.

Vitreous silica and a few other fully linked network glasses display negative thermal expansion coefficients over a limited temperature range (Figure 7.9). These glasses are very useful for applications requiring insulating materials which are highly resistant to thermal shock and which maintain their dimensions during thermal cycling. The negative thermal expansion coefficients are believed to result from the ability of the network to absorb lattice expansion through bending of bonds into the empty interstices of the structure.

Addition of modifier ions to silica fills the interstices, preventing bond bending, and hence increases the thermal expansion coefficient. The thermal expansion coefficients of binary alkali silicate glasses increase in

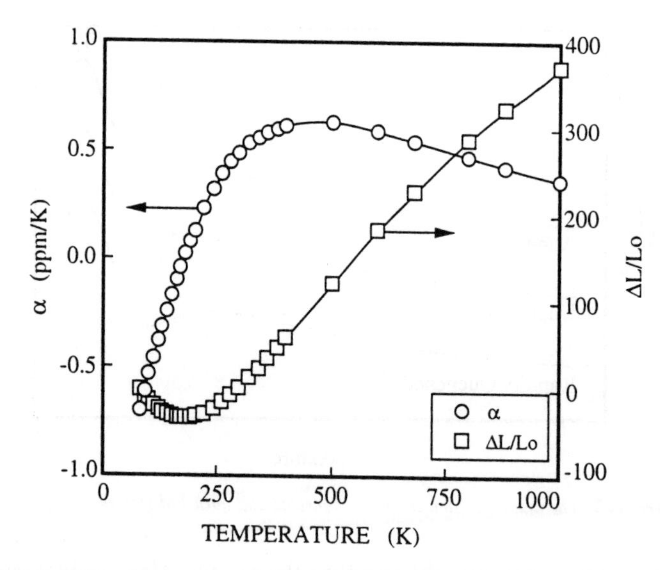

Figure 7.9 *Effect of temperature on the thermal expansion coefficient and specimen length of vitreous silica*

the order Li < Na < K. The thermal expansion coefficient is virtually independent of the existence of phase separation, increasing linearly with increasing alkali oxide content for all three oxides over the range from 0 to 25 mol % alkali oxide. Expansion factors can be derived by fitting the data to the expression

$$\alpha = 0.55 + x\alpha_x \tag{7.10}$$

where α is given in ppm K^{-1}, x is the mol % of the specified alkali oxide in the glass, and α_x is an expansion factor for that oxide. Values of α_x for Li_2O, Na_2O, and K_2O are 0.36, 0.48, and 0.58 ppm K^{-1} (mol %)$^{-1}$, respectively, for alkali silicate glasses in the temperature range from 300 to 400 °C. Values of α_x are a function of temperature and cannot be used outside the range for which they were derived. The increase in thermal expansion coefficient arises from the increase in non-bridging oxygen concentration, which increases the average asymmetry of the Si–O bond, and from the filling of interstices, which interferes with the bond bending mechanism responsible for the low thermal expansion of vitreous silica.

A small mixed-alkali effect occurs in the thermal expansion coefficient of alkali silicate glasses. The deviation from additivity due to this effect has been shown to increase with increasing total alkali oxide concentration and to be a function of the radius ratio of the larger to the smaller alkali ion present in the glass, with a maximum for the ratio for the sodium–potassium pair. The deviation from additivity is positive for radius ratios less than approximately 1.6, but is negative for larger radius ratios (Na–Cs and Li–Cs). Similar effects are found for mixed alkali borate and germanate glasses, where the deviations from additivity are also positive for mixtures of similar sized alkali ions, but are negative for mixtures of very different sized alkali ions.

Replacement of Na_2O by an alkaline earth oxide while maintaining a constant silica concentration decreases the thermal expansion coefficient of silicate glasses due to strengthening of the network by replacement of the low-field-strength sodium ion by the higher-field-strength divalent alkaline earth ions. The thermal expansion of sodium–alkaline earth–silicate glasses increases in the order Mg < Ca < Sr < Ba.

Additions of intermediate oxides (Al_2O_3, Ga_2O_3) to alkali silicate glasses decrease the thermal expansion coefficient by reducing the concentration of non-bridging oxygens and thus eliminating these highly asymmetric bonds. Gallium oxide has a smaller effect than that of alumina. Additions of other high field strength ions also tend to decrease the thermal expansion coefficient of silicate glasses, with values in the range of 2–6 ppm K^{-1} for glasses in the rare earth aluminosilicate and yttrium aluminosilicate systems.

Vitreous boric oxide has a large, very temperature dependent thermal expansion coefficient in the range of 15 ppm K^{-1}. The large thermal expansion coefficient is due to the two-dimensional nature of the structure of this glass, with weak bonds in the third dimension. Additions of alkali oxides initially decrease the thermal expansion coefficient (the *borate anomaly*), which passes through a minimum at ≈20 mol % alkali oxide and then increases monotonically to the glass formation limit for each system (Figure 7.10). The thermal expansion coefficient decreases in the order Cs > Rb > K > Na > Li, with a minimum value of 6 ppm K^{-1}. for a $20Li_2O–80B_2O_3$ glass. The initial decrease in thermal expansion coefficient is primarily due to the conversion of boron from triangular to tetrahedral coordination, which increases the connectivity of the structure. The reversal in thermal expansion coefficient at higher alkali oxide concentrations is due to changes in the intermediate structural units and to the eventual onset of non-bridging oxygen formation with increasing alkali oxide content.

Although vitreous germania has a structure very similar to that of vitreous silica, its thermal expansion coefficient is approximately an order of magnitude larger. Gas diffusion studies indicate that the free volume of the vitreous germania structure is much smaller than that of

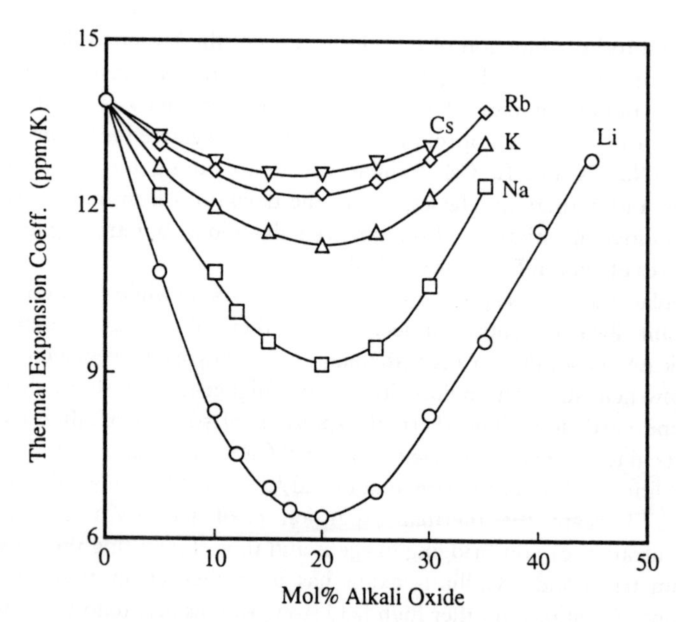

Figure 7.10 *Effect of composition on the thermal expansion coefficient of alkali borate glasses*

the vitreous silica structure. Apparently this difference in free volume restricts the bond bending process, which causes the low thermal expansion of vitreous silica.

Addition of alkali oxides to germania initially reduces the thermal expansion coefficient, which passes through a minimum at 2–5 mol % alkali oxide. Further additions of alkali oxides result in a continuous increase in the thermal expansion coefficient out to the limit of glass formation. The position of the minimum in the thermal expansion coefficient is near the low alkali *germanate anomaly* in viscosity and glass transformation temperature, which occurs at $\approx$2 mol % alkali oxide. No unusual behavior in the thermal expansion coefficient is found in the 15–20 mol % alkali oxide region, where the traditional germanate anomaly in density and refractive index occurs. Replacement of alkali oxides by alumina reduces the thermal expansion coefficient, but has little effect on the shape of the thermal expansion coefficient *versus* composition curve, which still displays a minimum at 2–5 mol % alkali oxide.

The thermal expansion coefficients of chalcogenide glasses are very large compared to those of oxide glasses. Vitreous selenium, for example, has a thermal expansion coefficient of $\approx$47 ppm K^{-1}, which decreases linearly with arsenic additions to 21 ppm K^{-1} for an As_2Se_3

glass. Arsenic sulfide glasses have even larger thermal expansion coefficients, with values as large as 72 ppm K^{-1} for an AsS_{10} glass.

Thermal expansion curves for amorphous metals closely resemble those of non-metallic glasses. Expansion coefficients for the composition $Fe_{40}Ni_{40}P_{14}B_6$ are almost identical in the amorphous and crystalline state, with values of 10.8 and 10.4 ppm K^{-1}, respectively. Similar results have been reported for other iron-based amorphous alloys. The thermal expansion coefficient is usually slightly greater for the alloy in the amorphous state than in the crystalline state. Glass transformation temperatures for these particular alloys lie in the range from 375 to 425 °C. The dilatometric softening temperature is often not detectable owing to crystallization before the sample reaches T_d.

Phase Separated Glasses

Phase separation has almost no effect on the thermal expansion coefficient of glasses. The measured thermal expansion coefficient is a volume average of the thermal expansion coefficients of the two phases present, in much the same way as the density.

On the other hand, neither T_g nor T_d is an averaged property. Each glass transformation will affect the curve independently of the existence of the other, so that each T_g may be observed. Softening of the sample, which determines the value of T_d, will be controlled by the more viscous phase if that phase is continuous, *i.e.*, if the microstructure consists of either a matrix of the more viscous phase with spheres of the lower viscosity phase, or if both phases are interconnected.

The thermal expansion curve of a phase separated sample containing two glassy phases may show two glass transitions, as is shown in Figure 7.11. This type of curve will be observed if the more viscous phase is continuous and if the immiscibility temperature for the glass lies above the T_g of the more viscous phase. Curves of this type have been observed for lead borate and barium silicate glasses, where the immiscibility temperature is greater than the T_g of either phase.

It is also possible for the thermal expansion curve of a phase separated glass to show only one glass transition (Figure 7.11). In this case, the curve may either closely resemble that of a homogeneous glass, which will occur if the more viscous phase occurs only in droplets, or may display a broad region of gradual softening, which will occur if the more viscous phase initially has a connected morphology. The former type of curve occurs for barium, sodium, or lithium silicate glasses containing modifier oxide concentrations beyond the spinodal limit, or for lead borate glasses containing only a small amount of PbO. Curves displaying

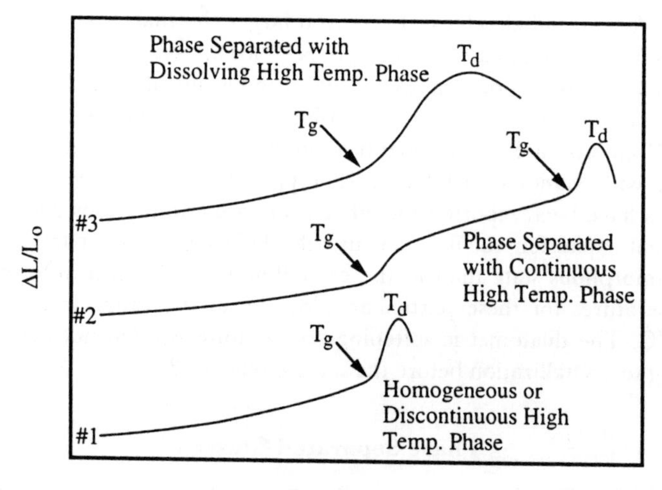

Figure 7.11 *Effect of phase separation on the thermal expansion curves of glasses*

a broad softening region are found for glasses where the rising sample temperature passes the immiscibility or spinodal limit, thus eliminating the connectivity of the higher viscosity phase, before the sample reaches the T_d appropriate for that phase. Curves of these types are observed for lithium and sodium silicate glasses containing smaller alkali oxide contents. In general, if the difference between T_g and T_d is more than $\approx$50 K, the sample is probably phase separated, with a continuous higher viscosity phase. If T_d minus T_g is less than 40 K, the sample is either homogeneous or the more viscous phase occurs only in droplet form.

Regardless of the overall shape of the thermal expansion curve, the T_g observed at the lower temperature, or if only one occurs, the lone T_g, will be that of the less viscous phase. It follow that, if two transitions are detected, the lower temperature will represent the T_g of the less viscous phase, while the higher temperature will indicate the T_g of the more viscous phase.

The relation between T_g and the composition of the lower viscosity phase suggests that thermal expansion measurements can be used to determine the tie lines in binary and ternary systems containing phase separated regions. Since all glasses on a tie-line have the same compositions for the two phases present, it follows that all bulk compositions lying on a given tie line will have the same T_g. Determination of the position of lines of constant T_g in a ternary phase separated region has proven to be the best method to date for defining the tie-lines in such a system. The same data can be used to determine the limits of immiscibility by determining the points where tie-lines end, *i.e.*, where the T_g ceases to be constant.

Thermal History Effects

Changes in fictive temperature have only a small effect on the thermal expansion coefficient for glasses at temperatures well below the transformation region. Heating a glass into the transformation region, however, allows relaxation to occur at rates which are comparable to the heating rates (3–5 K min^{-1}) used in these measurements. If we examine the curves shown in Figure 7.8, we find that reheating a quenched glass at a slower rate allows relaxation to occur before we reach the glass transformation temperature. The glass will actually contract at temperatures well below T_g. In extreme cases, the contraction can be so great that the sample will become shorter than its initial length. The determination of T_g from such a curve is no longer possible using the slope intercept method discussed previously.

A very different behavior is observed when reheating a well annealed glass, or one which was initially cooled at a rate less than the heating rate used in the thermal expansion measurement. In this case, the low fictive temperature of the glass and the corresponding high viscosity of the frozen structure allow the sample to pass the equilibrium line before relaxation occurs at a significant rate. Once relaxation begins to occur at a measurable rate, the sample will expand rapidly as it seeks to recover to the equilibrium volume line. Determination of T_g using this curve will yield a value in excess of that which would be determined for a glass which had been cooled at the same rate as the heating rate used during the expansion measurement.

Effect of Crystallization

Since the thermal expansion coefficient is a volume averaged function of the contributions of each of the phases present in a sample, formation of crystals with thermal expansion coefficients which are very different from the initial glass can radically alter the thermal expansion coefficient of the composite. Formation of crystals can also change the values of T_g and T_d by changing the composition of the residual glass and by preventing deformation of the sample under the push-rod load.

Commercial lithium aluminosilicate glass-ceramics provide excellent examples of such behavior. The initial glass used for production of transparent cookware, for example, has a thermal expansion coefficient of ≈ 4 ppm K^{-1}, $T_g \approx 730\,°C$, and $T_d \approx 760\,°C$. After processing, the thermal expansion coefficient is ≈ 0.5 ppm K^{-1} and T_g and T_d can no longer be detected on an expansion curve below 1000 °C. Heat treatment results in the formation of a lithium aluminosilicate crystal which has a very low thermal expansion coefficient. Removal of lithium from the residual glassy phase also decreases the thermal expansion coefficient of that phase, while simultaneously increasing the transformation and softening temperatures.

Crystallization of a commercial machinable glass-ceramic leads to a somewhat different behavior. This glass-ceramic contains only $\approx 50\%$ crystals. A glass transformation temperature at $\approx 460\,°C$ is clearly evident in the expansion curve of the glass-ceramic. This T_g is near that of the original glass. The dilatometric softening temperature, however, increases from $\approx 500\,°C$ for the base glass to over 900 °C for the glass-ceramic as a result of the interference with flow which results from the presence of the crystals.

The effect of crystallization on the thermal expansion behavior of a given glass also depends upon the identity of the crystalline phase formed. Heat treatment of the base glass used to produce one commercial lithium silicate glass-ceramic can yield either ≈ 33 vol % lithium metasilicate or ≈ 55 vol % lithium disilicate. Formation of lithium metasilicate will result in an increase in T_g from ≈ 480 to 620 °C, while T_d increases from 520 to 720 °C. Formation of lithium disilicate will yield a smaller increase in T_g to $\approx 570\,°C$, but a larger increase in T_d to $\approx 800\,°C$. The difference in composition of the residual glass due to formation of different crystalline phases accounts for the difference in the change in T_g, while the greater degree of crystallinity of the sample containing lithium disilicate accounts for the greater increase in T_d for that sample.

It is also possible for a sample to begin to crystallize during a thermal expansion measurement as the temperature approaches or just passes T_d. If this occurs, deformation of the sample may cease and the sample will begin to expand again. In other cases, a crystalline phase may melt, allowing a rapid decreases in viscosity and an abrupt softening of the sample.

SUMMARY

Density, molar volume, and thermal expansion coefficients are determined by the structure and bonding within a glass and are therefore strongly affected by changes in glass composition. Phase separation has little effect on these properties. The glass transformation and dilatometric softening temperatures are related to the viscosity of the melt and are strongly affected by both the composition of the glass and by the microstructure of phase separated glasses. All of these properties are altered by crystallization.

Chapter 8

Transport Properties

INTRODUCTION

A number of properties of glasses are controlled by the diffusion, or transport, of atoms or ions through the vitreous network. The electrical conductivity of almost all inorganic glasses containing monovalent ions is controlled by the diffusion of these monovalent ions under the influence of an external electric field. Chemical dissolution often begins with interdiffusion of alkali ions from glasses and proton species from the surrounding fluid. Ion exchange between alkali ions in the glass and a surrounding melt can be used to strengthen glasses by producing a compressive surface region. Other ion exchange processes are used to produce optical devices based on a graded refractive index from the surface to the bulk of the glass. Dielectric and mechanical losses are often due to movement of mobile ions under the influence of a reversing electric or stress field. Permeation of gases into and through glasses is controlled by the mobility of atoms or molecules through the vitreous network, as are many diffusion-controlled reaction processes.

FUNDAMENTALS OF DIFFUSION

The *diffusion coefficient*, or *diffusivity*, D, is defined by *Fick's First Law*,

$$J = -D\frac{\partial c}{\partial x} \tag{8.1}$$

where J is the flux of diffusing species and $\partial c/\partial x$ is the concentration gradient of the diffusing species in the direction x. The rate of change of concentration, c, with time at a given distance into the sample is given by *Fick's Second Law*, or

$$\frac{\partial c}{\partial t} = D\frac{\partial^2 c}{\partial x^2} \tag{8.2}$$

The diffusion coefficient of alkali ions in glasses is usually found by placing a thin layer of a source of a radioactive isotope of the ion on the surface of the sample, heating at a known temperature for a known time, t, and analyzing the concentration profile of the radioactive isotope in the glass using standard methods. The data for concentration *versus* distance are then fitted to the expression

$$c = \frac{Q}{\sqrt{\pi Dt}}\exp\left(-\frac{x^2}{4Dt}\right) \tag{8.3}$$

where Q is the concentration of the isotope per unit area placed on the glass surface and c is the concentration of the isotope at a distance x from the surface. Since Q and t are constants, we can obtain the diffusion coefficient from the slope of a plot of the log concentration *versus* x^2.

If the source of the diffusing species is a melt or a gas, the corresponding expression for the concentration profile is given by

$$c = C_0\mathrm{erfc}\left(\frac{x}{\sqrt{4Dt}}\right) \tag{8.4}$$

where C_0 is the constant surface concentration and erfc indicates the error function complement, which can be found in standard tables. The diffusion coefficient is now obtained from a best fit to the concentration profile obtained in exactly the same manner as was used for a thin layer source.

If the diffusing species is a gas such as helium, it is possible to expose one face of a plate of thickness L to a known pressure of that gas, while maintaining the other face at zero pressure. Under these conditions, a steady-state flow will be reached and Equation 8.1 can be rewritten as

$$J = -D\frac{\Delta c}{L} \tag{8.5}$$

where $\Delta c/L$ is the slope of the linear concentration gradient from the front to the rear surface. If we assume that the concentration of gas is given by *Henry's Law*, $c = SP$, where S is the *solubility coefficient* (or often simply the *solubility*) and P is the gas pressure, we can rewrite Equation 8.5 in the form

$$J = -D\frac{S(P_2 - P_1)}{L} \tag{8.6}$$

If we maintain a vacuum on the inside face of the sample, P_2 is zero, and P_1 equals the applied pressure, P. This expression can then be written as

$$\frac{JL}{P} = K = DS \tag{8.7}$$

where K is known as the *permeability coefficient*, or often simply the *permeability*. The permeability coefficient allows calculation of the steady-state flow rate of gas through a known area of a sample of known thickness when the sample is exposed to a known pressure of gas.

If the electrical conductivity of a substance is the result of field-induced diffusion of a single ionic species, the electrical conductivity, σ, of the substance is related to the diffusion coefficient, D, of that species *via* the Nernst–Einstein relation

$$\sigma = \frac{Z^2 F^2 Dc}{fRT} \tag{8.8}$$

where Z is the ionic charge, F is the Faraday constant, c is the concentration of the diffusing species in the glass, R is the gas constant, T is the temperature in K, and f is a constant which is experimentally determined and ranges from ≈ 0.2 to 1.0 for glasses. The constant f, which is sometimes termed the *Haven ratio*, has not been well characterized, with large variations among reported values for supposedly identical glasses.

Since diffusion is an activated process, it is not surprising to find that the temperature dependence of the diffusion coefficient is given by an expression of the form

$$D = D_0 \exp\left(\frac{-\Delta H_D}{RT}\right) \tag{8.9}$$

where D_0 is a constant and ΔH_D is the activation energy for diffusion. If we replace D in the Nernst-Einstein relation (Equation 8.8) by Equation 8.9 and combine the constants into a single term, we obtain the expression

$$\sigma T = \sigma_0 \exp\left(\frac{-\Delta H_D}{RT}\right) \tag{8.10}$$

which describes the temperature dependence of the ionic conductivity. In many cases, one finds that the use of Equation 8.10 to describe electrical conductivity is not quite correct and that a better fit to the data is obtained if the temperature is removed from the product σT, so that the data are represented by the expression

$$\sigma = \sigma_0 \exp\left(\frac{-\Delta H_D}{RT}\right) \tag{8.11}$$

where σ_0 and ΔH_D have different values from the corresponding terms in Equation 8.10. Since both forms of this expression are routinely found in the literature, one must be certain which expression has been used before attempting to calculate conductivities from listed values of σ_0 and ΔH_D. Care must also be used in extracting conductivities from graphs, which may be plotted as either $\log(\sigma)$ or $\log(\sigma T)$ as a function of reciprocal temperature.

Diffusion coefficients are also pressure and stress dependent. Measurements of the effect of pressure on electrical conductivity yield a linear relation, as expressed by

$$\sigma = \sigma_0 \exp\left(\frac{V^* P}{RT}\right) \tag{8.12}$$

where V^* is called the *activation volume* and P is the applied pressure. The constants σ_0 and V^* are independent of pressure at constant temperature. Changes in the volume of glasses produced by external pressure (compaction) are smaller than the changes in volume produced by changes in internal structure (fictive temperature), which creates an internal tension.

IONIC DIFFUSION

In theory, diffusion coefficients can be measured for any ion. In practice, however, most studies of ionic diffusion in glasses have been restricted to highly mobile ions which have a convenient radioactive isotope for use in tracer measurements. As a result, a majority of the data for ionic diffusion deals with sodium, with lesser amounts of data for potassium, rubidium, and cesium. Studies of lithium are very limited due to the lack of a radioactive isotope of lithium, while studies of divalent and other, more highly charged, ions are restricted by the very low mobilities of these ions compared to those of the monovalent ions.

Sodium diffusion coefficients in sodium silicate glasses increase with increasing sodium oxide concentration, as is shown in Figure 8.1. The scatter in the data is obviously quite large, with reported values for supposedly identical glasses varying by an order of magnitude in some cases. This scatter has been attributed to real differences among the

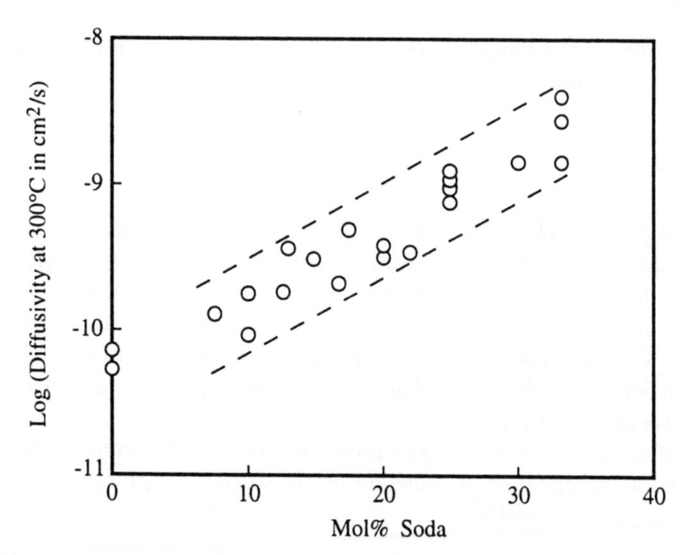

Figure 8.1 *Effect of sodium concentration on the diffusivity at 300 °C of sodium ions in sodium silicate glasses*

samples and not to experimental errors. It has been stated that differences in thermal history can alter sodium diffusivity by as much as an order of magnitude, while differences in hydroxyl content can alter sodium diffusivities by a factor of two. The increase in sodium diffusivity with increasing soda concentration is primarily due to a decrease in the activation energy for diffusion, with little systematic variation in the pre-exponential constant, D_0.

Diffusion data for other alkali ions are quite limited. If we consider potassium diffusion in potassium silicate glasses, we find trends similar to those for sodium in sodium silicate glasses, with increasing potassium diffusivities and decreasing activation energies for diffusion with increasing potassium oxide concentration in the glass. Data for rubidium and cesium diffusion in rubidium and cesium silicate glasses, respectively, are even more limited, but the same general trends appear to exist. If we compare the diffusivities of Na^+, K^+, Rb^+, and Cs^+ in R_2O–$3SiO_2$ glasses containing the appropriate alkali oxide, we find a small decrease in the diffusivity of the alkali ion in the order $Na^+ > K^+ > Rb^+ > Cs^+$. This difference, which is less than an order of magnitude at 400 °C, is smaller than one might expect from the large differences in ionic diameters of these ions.

Diffusion of an ion which is a primary component of a glass, as discussed above, is termed *self-diffusion*. We can also measure diffusion of an ion which is not present in the glass, *e.g.*, potassium diffusion in a sodium silicate glass. The results of such measurements for a set of R_2O–$3SiO_2$ glasses, where R is Na, K, Rb, or Cs, reveals that the diffusion coefficient is always greatest for the ion which is the component of the glass. The diffusivity of the impurity ions decreases as the difference between the size of the component ion and the foreign ion increases. The diffusivity of the alkali ions in a sodium silicate glass thus decreases in the order $Na^+ > K^+ > Rb^+ > Cs^+$, whereas that in a cesium silicate glasses decreases in the order $Cs^+ > Rb^+ > Na^+$. This relation between ionic diffusivity and glass composition gives rise to the *mixed-alkali effect* in conductivity, which will be discussed later in this chapter.

Addition of a divalent modifier such as Ca^{2+} to a sodium silicate glass decreases the diffusivity of Na^+ ions. The much less mobile divalent ions occupy interstices in the network and block the diffusion of the more mobile monovalent ions. This effect on diffusivity is at least partially responsible for the improvement in chemical durability of alkali silicate glasses which occurs when alkaline earth oxides are added to the composition.

Sodium diffusion in sodium germanate glasses behaves very differently from that found in sodium silicate glasses. The sodium diffusion coefficient in these glasses passes through a minimum for glasses containing between 10 and 20 mol % soda instead of increasing monotonically with increasing soda concentration, as in the sodium silicate glasses. This effect gives rise to very unusual behavior for the electrical conductivity of these glasses.

ION EXCHANGE

Ion exchange, or ionic *interdiffusion*, occurs when a glass containing one mobile ion, A, is exposed to a source of a different mobile ion, B. Ions from the glass diffuse out of the sample, while ions from the source diffuse into the sample. Since these ions have different sizes, their mobilities in the glass are different. The faster ion will tend to outrun the slower ion, which will cause an electric field to develop within the glass. This field will act to slow the faster ion and to accelerate the slower ion, until the fluxes of the two ions are identical. The overall process can be described by an interdiffusion coefficient, $\bar{D}$, which is given by

$$\bar{D} = \frac{D_A D_B}{D_A C_A + D_B C_B} \tag{8.13}$$

where D_A and D_B are the tracer diffusion coefficients of the ions A and B and C_A and C_B represent the fractional concentrations of the ions A and B. As the value of C_A approaches unity, the value of D approaches D_B, which is the diffusivity of the impurity ion.

Ion exchange can be used to alter the near surface properties of an existing glass. Recently, for example, exchange of sodium or other alkali from a glass with silver from $AgNO_3$ or other silver salts or from oxidized metallic silver films has been used to produce localized variations in refractive index. These variations can be used to guide light in glass plates (*planar waveguides*) or to produce lenses based on the principle of *graded refractive indices.*

Exchange of alkali ions can be used to strengthen glasses by producing a compressive layer in the near surface region. If exchange is carried out at temperatures well below T_g, very little stress relaxation will occur during the exchange process. Furthermore, if the ion in the glass is replaced by a larger ion from the external source, the difference in volumes of the two ions will result in a large compressive stress in the exchanged region. This process, which is termed *ion exchange strengthening,* or *chemical tempering,* can produce very high surface compressions of the order of 300–400 MPa. Most commercial ion exchange strengthening is based on replacement of sodium ions in the glass by potassium ions from a molten salt bath. This choice is based on both economics (the other alkalis are more expensive) and the fact that the sodium–potassium pair typically has a very favorable interdiffusion coefficient.

Special silicate glasses having high sodium diffusivities have been developed for ion exchange strengthening. These glasses have compositions containing roughly equimolar concentrations of soda and alumina, where both T_g and sodium mobilities are high. Commercial borosilicate glasses do not strengthen significantly due to their low soda contents, while relaxation during exchange prevents the development of large compressive stresses in soda–lime–silica glasses.

IONIC CONDUCTIVITY

Most oxide glasses, including silicates, borates, germanates, and most phosphates, are ionic conductors, while chalcogenide and some phosphate glasses are electronic conductors. Since the current carriers in ionically conducting glasses are ions, the electrical conductivity is closely related to the mobility of these ions. As a result, glasses which contain significant concentrations of monovalent ions are poor insulators, while glasses which are free of monovalent ions are excellent insulators.

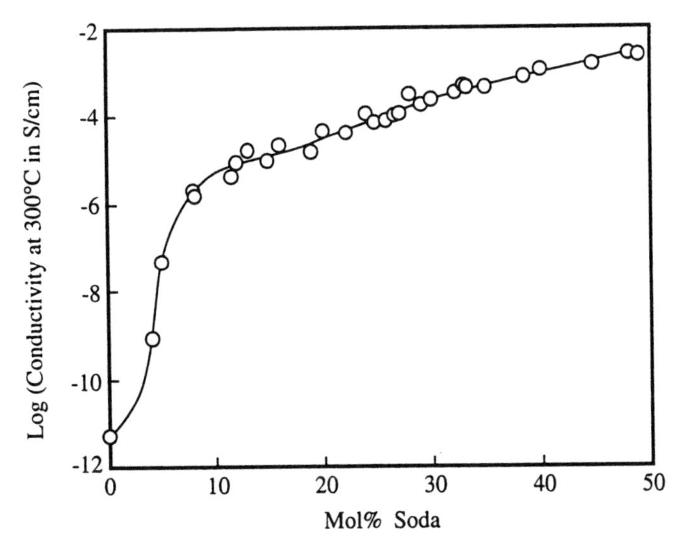

Figure 8.2 *Effect of glass composition on the electrical conductivity at 300 °C of sodium silicate glasses*

Compositional Effects

The electrical conductivity of a glass depends not only upon the mobility of the current carrier, but also on the concentration of carriers (see Equation 8.8). As a result, the conductivity of sodium silicate glasses increases more rapidly with increasing soda concentration than if it were only controlled by the sodium diffusion coefficient (Figure 8.2). The conductivity of other alkali silicate glasses also increases with increasing alkali oxide content, with a very small decrease in conductivity for corresponding glasses containing the same alkali oxide concentration in the order Na > K > Rb > Cs.

Glasses containing two or more alkali oxides display the *mixed-alkali effect,* as shown in Figure 8.3 for sodium–potassium silicate glasses. Although the total alkali oxide content of these glasses is constant, the electrical conductivity passes through a large minimum as the ratio of sodium to total alkali oxide concentration varies from zero to one. Simple additivity arguments would suggest that the conductivity of these glasses would vary linearly with this ratio. Similar behavior is found in many systems, including simple alkali borate and germanate, alkali aluminosilicate, alkali galliosilicate, and other glassforming systems. Although it is sometimes stated that non-bridging oxygens are required for the existence of the mixed-alkali effect, this contention is erroneous.

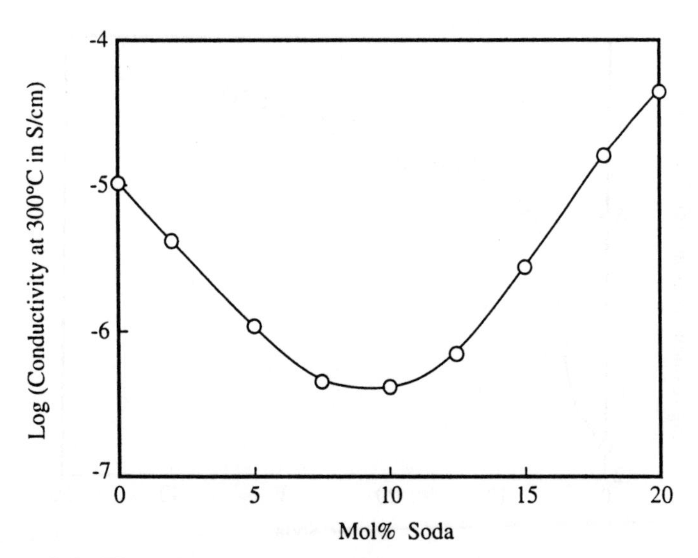

Figure 8.3 *The mixed alkali effect on the electrical conductivity at 300 °C of sodium–potassium silicate glasses containing 20 mol % total alkali oxide*
(Data supplied by J. J. Noonan)

The mechanism leading to this behavior has been the subject of many studies, with at least 10 suggested explanations. Examination of the diffusion coefficients of sodium and potassium in such a series of glasses reveals that each coefficient decreases monotonically as the concentration of the other alkali ion increases. The cross-over in the diffusivities occurs near the minimum in conductivity and the calculated conductivity using the Nernst–Einstein relation is near the experimental value. The cause of the trends in the diffusivity remains uncertain at this time.

Addition of an alkaline earth oxide to a glass containing alkali ions also decreases the electrical conductivity as well as the diffusivity of the alkali ion. This behavior is usually explained as due to the blocking effect of the immobile divalent ion, which occupies interstices that can therefore no longer be used for alkali migration. It may well be, however, that the effect of the divalent ion is simply another manifestation of the mechanism causing the mixed-alkali effect.

The electrical conductivity of alkali aluminosilicate and galliosilicate glasses is rather interesting (Figure 8.4). The isothermal conductivity of these glasses initially decreases as the intermediate oxide is added to the glass, even if the alkali oxide concentration is held constant. The conductivity passes through a minimum at an intermediate to alkali ratio between 0.2 and 0.6 and then rises to a maximum at a ratio of 1.0–1.1,

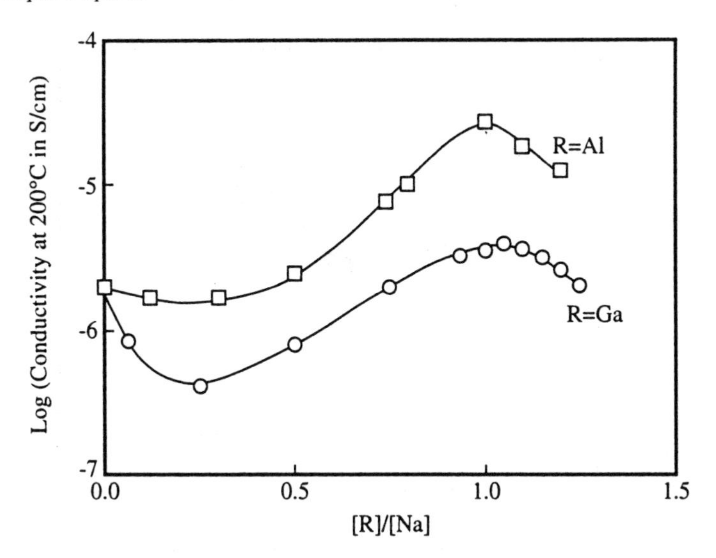

Figure 8.4 *Electrical conductivity at 200 °C of sodium galliosilicate and sodium aluminosilicate glasses*
(Data supplied by J. C. Lapp)

after which it decreases steadily until the glass formation limit is reached. This behavior has been attributed to the existence of two different alkali sites in these glasses: the alkali-NBO sites of intermediate-free glasses and the $R^+(MO_{4/2})^-$, where R is an alkali ion and M is either aluminum or gallium, sites which form in the glasses containing intermediate oxides. The conductivity/composition curve can be explained if one assumes that alkali ions do not freely move between these two sites. The initial addition of the intermediate oxide reduces the concentration of the first type of sites, which would reduce the mobility of the associated alkali ions. Since the concentration of the second type of sites is small, those alkali ions are effectively immobilized. Further increases in the concentration of the second type of site allow alkali ion movement between those sites, which counters the decrease in conductivity resulting from the decrease in the alkali-NBO sites. If the alkali associated with the intermediate-oxygen tetrahedra are actually more mobile than those associated with NBO, the overall result would be an increase in conductivity as the intermediate oxide content increases beyond some minimum value. The eventual maximum at an intermediate to alkali ratio of 1.0–1.1 is due to the undefined structural change which occurs in that compositional region.

The electrical conductivity of alkali germanate glasses behaves very

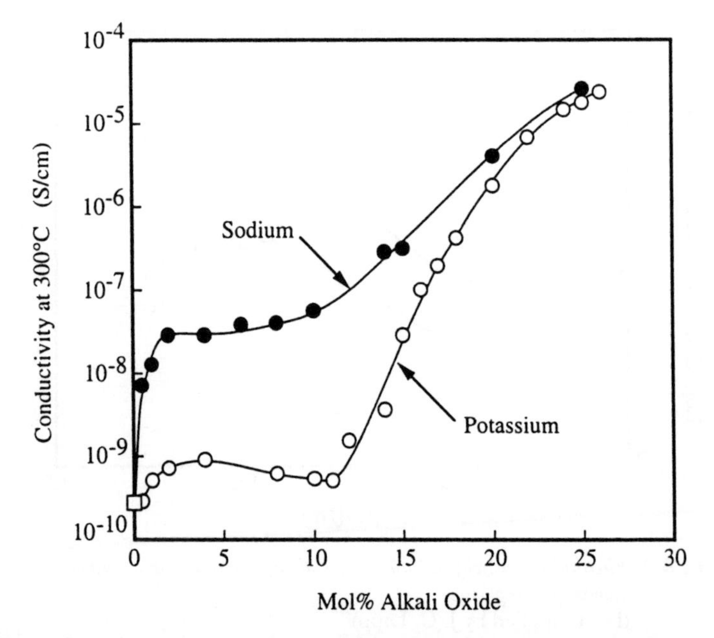

Figure 8.5 *Electrical conductivity at 300 °C of sodium and potassium germanate glasses*

differently from that of the alkali silicate and borate glasses (Figure 8.5). The conductivity of the lithium and sodium germanate glasses increases very rapidly when even a small amount of either alkali oxide is added to germania. This initial large increase in conductivity is not observed for potassium, rubidium, or cesium germanate glasses, where the electrical conductivity initially only increases by a small amount, after which it decreases and passes through a small minimum at 12 mol % alkali oxide, and then increases to values near those of glasses containing lithium or sodium oxide at alkali oxide concentrations exceeding 25 mol %. This behavior can be understood if one realizes that the unusual behavior of the density of these glasses affects the alkali concentrations in a very non-linear manner, which is not the case for alkali silicate glasses, and that the diffusivities of alkali ions in germanate glasses pass through a minimum with increasing alkali oxide concentration.

Although the electrical conductivity of most oxide glasses is due to cation migration, the electrical conductivity of many glasses containing halides, either as partial or complete replacements for oxygen, is due to anion migration. Anion conductivity in glasses was first reported for lead halosilicate glasses containing less than 50 mol % silica. The conductivity of these glasses increases by several orders of magnitude as PbX_2, where X is any halide, replaces PbO. Fluorine is much more mobile than the other halide ions, with conductivity increasing in the order I < Br < Cl < F for glasses containing the same concentrations of lead halide. These glasses have also been shown to display an effect, termed the *mixed-halide effect*, or the *mixed-anion effect*, analogous to the mixed-alkali effect if they contain two different halides.

Many other systems have subsequently been found to display anion conductivity. The electrical conductivity of lead fluoroborate and fluorogermanate glasses is similar to that of the lead fluorosilicate glasses. Calcium fluoroaluminate glasses also conduct by fluorine migration. Fluorophosphate glasses have also proven to be good anionic conductors. All of these systems are free of alkali ions.

The electrical conductivity of glasses containing both alkali ions and halides is only slightly affected by the replacement of oxygen by fluorine. While a deviation from additivity is found, for example, for sodium fluoroborate glasses as NaF replaces Na_2O, the effect is very small compared to the mixed-alkali effect which would be expected for glasses containing so much alkali.

Not surprisingly, heavy metal fluoride glasses are also anion conductors, with moderate electrical conductivities. It is surprising, however, to find that the conductivities of these glasses are typically considerably less than those of the lead fluorosilicate, fluoroborate, and fluorogermanate glasses, which contain considerably less fluorine. It is also interesting to find that the addition of sodium or lithium to heavy metal fluoride glasses decreases their electrical conductivity instead of increasing it. These effects are not clearly understood at this time.

Activation Energy for Electrical Conductivity

The activation energy for diffusion actually represents the sum of two independent terms. Diffusion requires that the bond between the ion and its immediate surroundings be broken before the ion can move to a new site. Movement from one site to another requires passage of the ion through the doorways between adjacent interstices. The overall process is frequently described using the concepts proposed by *Anderson and Stuart* in 1954. If we call the energy required to break the local bond the *electrostatic binding energy*, E_b, and the energy necessary to move the ion by straining a doorway the *strain energy*, E_s we can write the expression

$$E_\sigma = E_b + E_s \tag{8.14}$$

where E_σ is the activation energy for conduction. Anderson and Stuart further suggested that E_b can be approximated by the expression

$$E_b = \frac{\beta z z_0 e^2}{\gamma(r + r_0)} \tag{8.15}$$

where β is a factor which accounts for the distance between neighboring sites, z and r are the charge and radius of the cation, respectively, z_0 and r_0 are the charge and radius of the anion, respectively, e is the electronic charge, and γ is called the covalency parameter. The covalency parameter accounts for the deformability of the anion and is found experimentally to be equal to the dielectric constant of the glass. The strain energy can similarly be approximated by the expression

$$E_s = 4\pi G r_d (r - r_d)^2 \tag{8.16}$$

where G is the shear modulus of the glass and r_d is the diameter of the doorway between adjacent interstices.

A number of variations for E_s have been offered since the original Anderson and Stuart model was first proposed. While these models provide a better approximation of the experimental activation energies and correct some faulty predictions based on Equation 8.16, the basic concept of an activation energy for conduction consisting of both electrostatic and strain energy terms has remained unchallenged.

Effect of Phase Separation on Electrical Conductivity

The morphology of a sample can have a large effect on the electrical conductivity. If the alkali ions are isolated within spheres distributed in a matrix of a much lower conductivity phase, the dc conductivity will be found to be much lower than anticipated from the bulk glass composition. A large, discontinuous increase in conductivity will occur when these isolated spheres become connected, *i.e.*, the microstructure contains a connected high conductivity phase. This effect is shown in Figure 8.6 for the lithium silicate glasses, where the microstructure of any glass containing less than 8 mol % Li_2O is composed of lithium-rich spheres in a lithium-poor matrix. The change to an interconnected microstructure in the range from 8 to 10 mol % lithia is accompanied by a sharp increase in electrical conductivity. The activation energy for conduction, which is that of the phase controlling the overall conductivity of the glass, decreases sharply at the same lithia concentration. The electrical conductivity of glasses with a continuous high-alkali content phase will increase slightly with increasing volume fraction of that phase, as seen in the region from 10 to 33 mol % lithia.

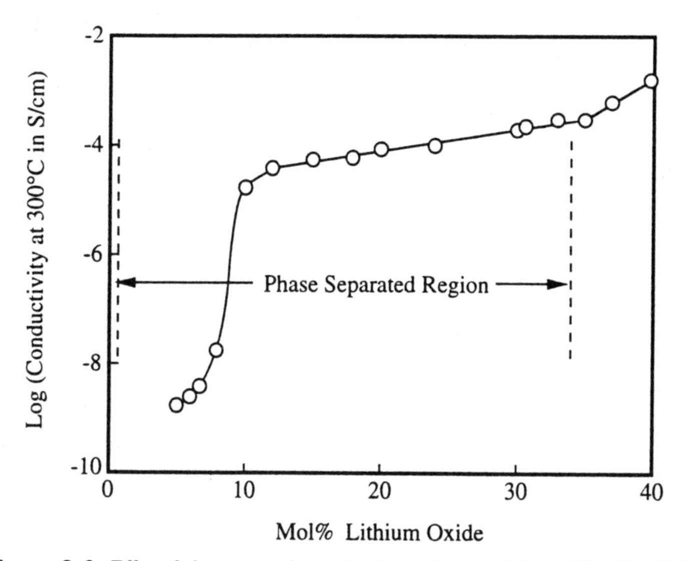

Figure 8.6 *Effect of phase separation on the electrical conductivity at 300 °C of lithium silicate glasses*
(Data supplied by B. M. Wright)

Effect of Thermal History on Electrical Conductivity

If we apply the Anderson and Stuart model to a consideration of the effect of thermal history on electrical conductivity, we would reason that alkali ions should require less energy to move through a structure with large doorways than through one with small doorways. If we further argue that doorway diameter and density of the glass are related, we will predict that the energy required for ion motion through a glass should increase as the density increases. Since an increase in activation energy will make movement more difficult, it follows that the conductivity of a glass should decrease as the density increases. In other words, the conductivity of a quenched glass should be greater than that of the same glass after annealing or following a slow cooling from the melt. This behavior is found for all common glasses.

Effect of Crystallization on Electrical Conductivity

The effect of crystal formation on electrical conductivity will depend largely upon the composition of the crystalline phases formed. If this phase removes alkali ions from the residual glass, the conductivity of the composite will usually decrease, while formation of an alkali-free crystalline phase may lead to an increase in conductivity. Common glass-ceramics based on either lithium silicate or aluminosilicate crystalline phases have conductivities which are several orders of magnitude less than that of the base glass before crystallization. Formation of quartz or other phases of silica, however, can increase the conductivity of the composite by increasing the conductivity of the residual glass, even though the amount of that glass is only a small volume fraction of the composite.

CHEMICAL DURABILITY

Although commercial silicate glasses are used for containers for liquids, one should not assume that glasses in general are not soluble in aqueous solutions. Many, if not most, non-silicate glasses are very susceptible to dissolution in water, which often limits their utility for any application involving contact with either aqueous liquids or water vapor. Even those glasses which exhibit excellent chemical durability in water may be readily dissolved if exposed to very high or low pH solutions or to a reagent such as hydrofluoric acid (HF), which specifically attacks the network bonds.

A number of processes have been observed during the dissolution of glasses in aqueous solutions. If the glass contains any alkali or other highly mobile ions, *ion exchange* between these ions and protonic species (probably hydronium, or H_3O^+ ions) from the liquid can occur. The liquid may also directly attack the network bonds in such a manner that the concentration ratios of the components in the liquid are identical to those in the glass. This process is known as *congruent dissolution* and can occur for any glass. While ion exchange can occur in the presence of either liquid or water vapor, congruent dissolution requires the presence of a liquid. Finally, layers of reaction products can form on the surface of the glass, which may influence the subsequent rate of dissolution of the material.

Since most common commercial silicate glasses contain alkali ions, the initial step in their dissolution usually involves ion exchange between these ions and protonic species from the liquid. The interdiffusion coefficient for the alkali ion and the protonic species is given by Equation 8.13, which can be rewritten in the form

$$\bar{D} = \frac{D_a}{1 + bN_a} \tag{8.17}$$

where D_a is the diffusion coefficient for the alkali ion, N_a is the ionic fraction of the alkali ions relative to the total concentration of mobile species, and b is a constant given by the expression

$$b = \left(\frac{D_a}{D_H}\right) - 1 \tag{8.18}$$

where D_H is the diffusion coefficient for the protonic species. This model has been found to yield good agreement for many glasses containing lithium or sodium ions, where the diffusion coefficients for the alkali ions agree quite well with values obtained by other methods.

Since the ion exchange process is diffusion controlled, it is not surprising to find that the depth of hydrogen penetration into glasses increases with the square root of time of exposure to the source of hydrogen. The alkali concentration in the surrounding solution also increases with the square root of time.

Congruent dissolution of the glass occurs simultaneously with ion exchange. Since congruent dissolution occurs at a constant rate, while ion exchange is proportional to the square root of time, congruent dissolution will dominate the dissolution process at longer times and the thickness of the exchanged layer will become independent of time. If one characterizes the chemical durability by a weight loss measurement, one thus finds that the initial rate of weight loss varies with the square root of time, with a change to a linear loss with time at longer times.

Solution conditions strongly influence the rate of dissolution of glasses. If the solution volume to sample area ratio is very large, or if the solution is constantly replaced, the solution remains dilute throughout the process and the pH remains constant. On the other hand, if the solution volume to sample area ratio is very small, ion exchange of alkali from the glass produces a rapid increase in the pH of the solution. This increase in pH increases the solubility of silica in the solution and thus increases the dissolution of the glass. If the glass is free of alkali or alkaline earth ions, the pH will not increase and the solution can become saturated in silica at low levels of silica concentration, which will cause the dissolution rate to decrease or even cease.

Extremes of pH usually result in rapid dissolution of glasses. In general, silicate glasses begin to dissolve rapidly at pH levels greater than 9 or less than 1. The rate of congruent dissolution becomes so great that the ion exchange process is no longer important and the leached layer thickness becomes negligible. Since the attack occurs directly on the Si–O bonds, differences in durability become small at these extremes of pH unless the glass contains other elements which alter the dissolution mechanism. Additions of alumina, for example, which typically improve durability in neutral solutions, result in very rapid dissolution in highly acidic solutions due to attack at the Al–O bonds. Zirconia is often added to glasses to improve their durability at high pH.

Changes in the morphology of phase separated glasses can alter their durability by orders of magnitude. Common alkali borosilicate glasses consist of two phases of very different compositions: an alkali borate phase and a silica-rich phase. While the alkali borate phase has a very low durability in even weak acids, the silica-rich phase has a relatively high durability in such solutions. If the alkali borate phase exists only as isolated droplets distributed throughout a silica-rich matrix, the durability of bulk samples will be determined by that of the silica-rich phase since the solution will not come in contact with the low-durability phase. If, however, the glass is subjected to a heat treatment which results in the formation of an interconnected morphology, continuous filaments of the low-durability alkali borate phase will exist throughout the material. Exposure of the heat treated glass to a weak acid can now result in leaching of the low-durability phase from the material, leaving a porous silica-rich glass.

Leaching of phase separated borosilicate glasses serves as the basis for production of Vycor® glass by Corning. This glass is sold in both the as-leached state, which contains ≈35% connected porosity with an average pore diameter in the range of a few nanometers, and in the consolidated state, which is obtained by heat treating the leached glass at temperatures ≥1000 °C. While the effect of morphology on the chemical durability of this particular glass has been used to produce a commercial product, it is also possible that inadvertent changes in the thermal history of otherwise durable glasses can seriously degrade their usefulness. One should also remember that data obtained from bulk samples of phase separated glasses may be very different from data obtained from finely ground samples, where exposure of otherwise isolated low durability spheres is greatly increased.

Crystallization can alter the durability of glasses by altering the chemical composition of the phases present. Crystallization often removes mobile ions such as lithium from the glass and places them into very durable crystals. In this case, the durability of the crystallized material is usually superior to that of the base glass. It is possible, however, for the crystals to be less durable than the base glass, in which case the final material may be subject to leaching by dissolution of the crystalline phase.

WEATHERING

The term *weathering* is often used to refer specifically to the interaction of glasses with water vapor, as opposed to the term *chemical durability*, which usually refers to the interaction with liquids. The quantity of water available at a glass surface exposed to water vapor is insufficient for significant congruent dissolution of the network. Ion exchange between the mobile ions in the glass and the adsorbed water molecules can still occur. Since no liquid exists, the ions which diffuse from the glass remain on the surface of the sample, where they can react with the surrounding atmosphere. Initially, hydroxides of alkali and alkaline earths form at the surface *via* reactions such as

$$Na^+ + 2H_2O \rightarrow H_3O^+ + NaOH \tag{8.19}$$

These hydroxides then react with carbon dioxide from the atmosphere to form carbonates, as in the reaction

$$2NaOH + CO_2 \rightarrow Na_2CO_3 + H_2O \tag{8.20}$$

or the reaction

$$Ca(OH)_2 + CO_2 \rightarrow CaCO_3 + H_2O \tag{8.21}$$

These carbonates, which occur as discrete particles on the surface of the glass, scatter light and give the glass a frosted appearance. Since these particles are tightly bonded to the glass surface, attempts to remove them result in permanent damage to the surface.

Another mechanism can add to the damage caused by weathering. If the humidity surrounding the glass surface cycles over a wide range, the hydrated, leached layer will vary in water content. A reduction in humidity will result in dewatering of this layer, which is associated with volume shrinkage. Since the leached layer is bonded to the unleached bulk glass substrate, which does not shrink, stresses will develop at the interface between the leached layer and the bulk glass. These stresses may result in peeling of the leached layer from the substrate, which can leave craters in the glass surface. These craters cause significant light scattering and seriously degrade the optical appearance of the glass.

Since weathering is a surface phenomenon, surface treatments can significantly alter the rate of weathering of otherwise identical glasses. Reduction in alkali concentration by treatment with SO_2 will produce a dealkalized near-surface region and improve weathering resistance. An extreme example of this effect can be found in common commercial soda–lime–silica glasses produced using the float process. Glasses produced by this process have a tin-rich surface due to 'floating' of the molten glass on molten tin during the formation of sheets of glass. Some diffusion of tin into the glass surface occurs during this process. Since the other surface of the glass was not in contact with the molten tin, it has a much lower tin content. (Some tin still enters this surface from the vapor above the molten tin bath.) Exposure of float glasses to water vapor clearly show the dramatic reduction in weathering of the tin-rich surface as opposed to the tin-poor surface.

GAS PERMEATION AND DIFFUSION

A number of gases can permeate through glasses at rates which can have serious consequences in many practical applications. Helium can readily permeate through many glasses used for vacuum tubes. Hydrogen permeation can result in the coloration of glasses by the reduction of ions to a lower valence or to the metallic state and by reaction with optical defects. Oxygen permeating through the thin wall of an electric lamp can react with the filament material, causing failure of the bulb.

Experimental results indicate that the permeability of a given glass to various gases decreases as the atomic or molecular diameter of the diffusing species increases. For vitreous silica, for example, gas permeability has been found to decrease in the order He > H_2 > Ne > N_2 > O_2 > Ar > Kr (Figure 8.7). Since the permeability of gases other than He, Ne, and hydrogen isotopes has not been measured for other glasses, the generality of this trend cannot be confirmed, but the trend for these three gases is always the same, *i.e.* He > H_2 > Ne.

The temperature dependence of gas permeability in glasses has been the subject of some controversy. The temperature dependence of the permeability, K, can be described by an expression similar to Equation 8.9, or

$$K = K_0 T^n \exp\left(\frac{-E_k}{RT}\right) \tag{8.22}$$

where K_0 is a constant, T is the temperature in K, n is a constant which has been proposed to be equal to 0, 0.5, or 1.0 in various papers, and E_k is the activation energy for permeation. The actual values of K_0 and E_k will depend upon the value of n used in fitting the experimental data. Extensive studies of helium permeation in vitreous silica have established that n = 1.0 for that particular combination of gas and glass. Unfortunately, studies covering a sufficiently wide temperature range to differ-

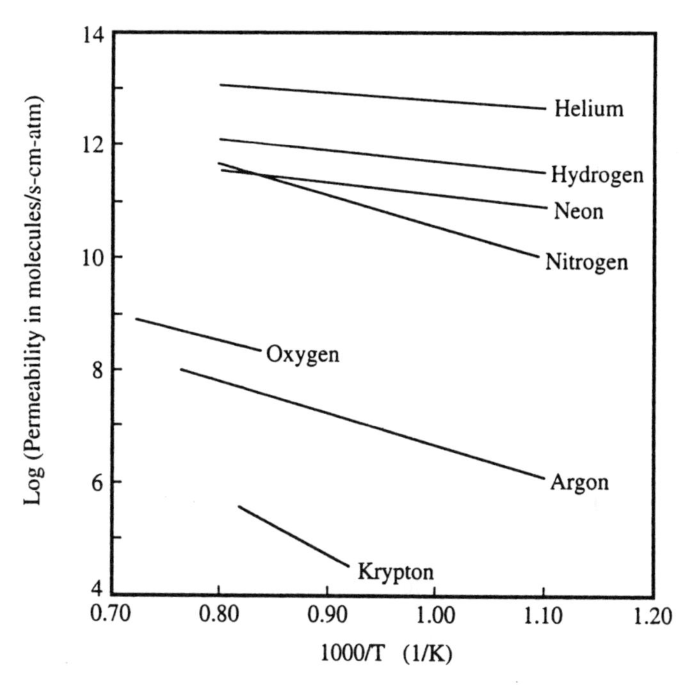

Figure 8.7 *Temperature dependence of the permeability of several gases through vitreous silica*

entiate values of n clearly do not exist for other gas/glass combinations. As a result, results of permeation studies for K_0 and E_k are reported using Equation 8.22 with either $n = 0$ or $n = 1.0$ in different papers. Care must be used to determine which form of Equation 8.22 was used in calculating these constants from the data before using the results for further calculations.

Gas permeability varies linearly with the partial pressure of the diffusing gas for all gases, for pressures up to many atmospheres. The diffusivity of the gas is independent of pressure for modest pressures. At high pressures, non-linear effects occur due to the thermodynamic requirement that gas fugacity be considered in place of gas pressure and due to the onset of saturation effects caused by filling of the interstices in the vitreous network by dissolved gas atoms or molecules.

The effect of glass composition on helium permeation has been studied for a wide variety of oxide glasses, including silicate, borate, germanate, and phosphate compositions. In general, helium permeability decreases in silicate glasses as the concentration of modifier ions increases. The modifier ions are believed to occupy the interstitial sites in

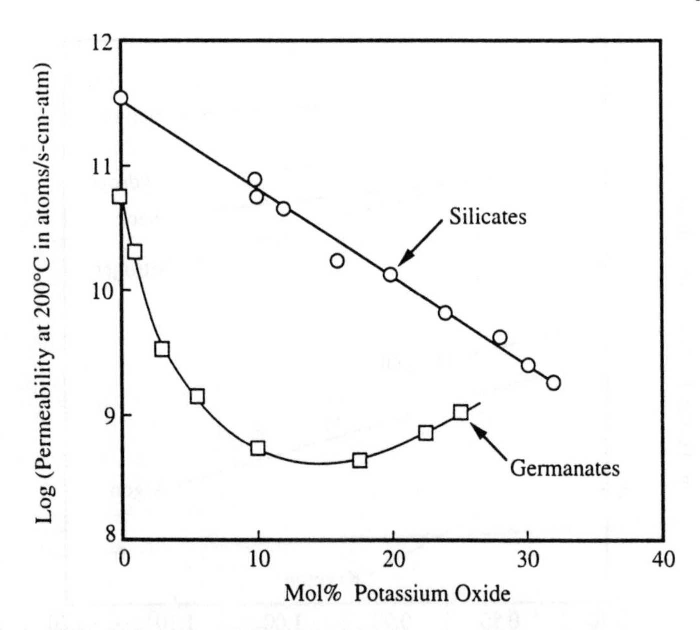

Figure 8.8 *Helium permeability at 200 °C in potassium silicate and potassium germanate glasses*

the network, blocking diffusion paths for the small helium atoms. Since lithium ions are believed to cause a contraction of the interstices, lithium oxide proves to be very effective at reducing helium permeability. Sodium ions, which occupy similar interstices but do not cause a contraction of the structure, are somewhat less effective at reducing permeability. Potassium, rubidium, and cesium ions, which actually expand the network, still fill interstices and reduce helium permeation, but are even less effective than sodium ions. As a result, the helium permeability of alkali silicate glasses decreases in the order Cs = Rb = K > Na > Li for glasses containing equal concentrations of the various alkali oxides. A plot of log permeability *versus* mol % modifier oxide yields a straight line for many of these systems (Figure 8.8), indicating that K_0 is only slightly affected by glass composition, while E_k increases linearly with increasing mol % of the modifier oxide.

Alkaline earth ions are also very effective in blocking the diffusion of helium atoms, with a similar trend toward decreasing helium permeability in the order Ba > Sr > Ca > Mg for glasses having otherwise identical compositions. The least permeable traditional commercial glasses are based on alkaline earth aluminosilicate compositions.

The relations between glass composition and helium permeability are much more complex for non-silicate glasses. Helium permeability still decreases with additions of modifier oxides to the pure glassforming oxide, but the permeability no longer decreases in a simple fashion with increases in modifier oxide concentration. The helium permeability of alkali germanate glasses (Figure 8.8), in fact, passes through a minimum and then increases with increasing alkali content. In general, the trend toward decreasing permeability in the order Cs > Rb > K > Na > Li is observed for these systems just as in the alkali silicate glasses.

Phase separation has a major effect on helium permeation in oxide glasses. The permeabilities of the two phases often differ by several orders of magnitude. Since the gas atoms will diffuse much more rapidly through a modifier-free glass, diffusion will occur through the modifier-free phase wherever possible. The presence of droplets of a modifier-rich phase has little effect on gas permeation. Any change in composition or thermal history, however, which disrupts the connectivity of the modifier-free phase will result in an abrupt decrease of a factor of 100 or more in gas permeability. Changes in connectivity of the alkali-rich phase, which dramatically alter other diffusion-controlled properties such as electrical conductivity and chemical durability, have virtually no effect on gas permeation.

At first consideration, it is somewhat surprising to learn that crystallization usually increases gas permeability in glasses. It seems logical to assume that replacement of a permeable glass by impermeable crystals would decrease the permeability of the composite material. In most cases, however, the formation of crystals removes modifier ions from the glassy phase, resulting in an increase of orders of magnitude in the permeability of the remaining glass. The exponential increase in permeability of the residual glass more than offsets the geometric decrease in permeability expected from the presence of the crystals, so that the overall permeability increases.

Organic glasses are also quite permeable to gases. The very open structures of many of these glasses results in very high permeabilities, with little difference in permeability with the identity of the permeating species. Other inorganic glasses have lower permeabilities and considerably larger differences in permeability among the gases. Gas permeabilities increase as the temperature passes through the glass transformation regions, where relaxation can expedite atomic and molecular diffusion. Since the glass transformation temperature of many organic glasses occurs below room temperature, the samples studied are actually rubbers, which have higher permeabilities than the corresponding glasses at lower temperatures. As a result, the room temperature permeability of many organic materials increases as the value of T_g decreases.

Amorphous metals do not have the high free volumes found in oxide and organic glasses. Their structures are so dense that the inert gases cannot pass through the doorways between interstices. Since hydrogen molecules dissociate on absorption into metals, they can diffuse as very small protons through these materials. The permeability of amorphous metals, just as for crystalline metals, appears to be primarily controlled by the nature of oxide films on their surface.

DIFFUSION-CONTROLLED REACTIONS

Many reactions in glasses are controlled by the diffusion of atoms, molecules, or ions from the surrounding environment into the glass. Since an extensive discussion of these reactions is beyond the scope of this book, only one typical example of a *diffusion-controlled* process will be presented here.

A large number of reactions occur between hydrogen gas and various reactive sites in glasses. These reactions include the reduction of iron from the iron(III) to the iron(II) state, similar reactions involving tin, chromium, manganese and other transition and rare earth ions, reduction of many ions, including silver, copper, gold, arsenic, lead, antimony, and bismuth to the atomic state, which is followed by agglomeration of the metal atoms to form colloids, and reactions with defects induced by radiation or by other extreme treatments of glasses.

All of these processes can be described by a diffusion-controlled model originally derived to explain the tarnishing of metals and hence commonly called the *tarnishing model.* The derivation of this model is based on the assumptions that (a) the reaction site is immobile, (b) the concentration of reaction sites is independent of time and temperature in the absence of the tarnishing reaction, and (c) the reaction rate is very fast compared to the diffusion time. Under these conditions, a sharp interface will exist between the reacted and unreacted portions of the sample. Measurements of the overall rate can be based on either the growth of the reacted layer, which can easily be measured for colloid formation, where the reacted zone is colored, or by measurement of the change in concentration of the reaction site species, which may be carried out by use of some spectroscopic technique. In the first case, the thickness of the reacted layer is given by the expression

$$X = \sqrt{\frac{2C_g D_g t}{C_x}} \tag{8.23}$$

where C_g is the concentration of dissolved gas at the outer surface of the sample, D_g is the diffusion coefficient for the gas in the sample, t is the elapsed time since exposure to the gas, and C_x is the reaction site concentration. If the solubility of gas obeys Henry's law, then $C_g = SP$, where S is the solubility of the gas in the sample and P is the gas pressure. Furthermore, since the permeability is equal to the product of the diffusion coefficient and the solubility, then $K = SD_g$. After substituting these relations into Equation 8.23, we obtain the expression

$$X = \sqrt{\frac{2KPt}{C_x}} \tag{8.24}$$

If the experimental conditions prevent direct measurement of the layer thickness, we can convert Equation 8.24 into a form appropriate for measurements of the average concentration of the reacted species in the sample. Under these conditions, for samples exposed to gas from both surfaces, one can show that the concentration of the reacted species, C, is given by the expression

$$\frac{(C - C_i)}{(C_f - C_i)} = \sqrt{\frac{8KPt}{C_x L^2}} \tag{8.25}$$

where C_i is the initial concentration of the reacted species, C_f is the final concentration of that species, and L is the sample thickness. Other forms of this expression can be derived for other experimental conditions, but all expressions will indicate that the thickness of the reacted zone will increase with the square root of time and that the overall rate is controlled by the permeability of the gas through the material. It follows that the term 'diffusion-controlled' is actually a misnomer and that these processes are actually *permeation-controlled* when dealing with reactions between gases and solids.

SUMMARY

Diffusion of atoms, molecules, and ions controls many processes in glasses, including ionic diffusion, ion exchange, electrical conduction, chemical durability, gas permeation, and permeation-controlled reactions. Since the mechanisms underlying all of these processes are based on similar principles, a fundamental understanding of diffusion phenomena serves as the basis for understanding all diffusion-controlled properties of glasses.

Chapter 9

Mechanical Properties

INTRODUCTION

Glasses are brittle materials. As a result, their fracture behavior is usually determined by environmental factors and not by the inherent strength of the bonds forming the vitreous network. The fracture strength of glasses varies with prior surface treatment, chemical environment, and the method used to measure the strength. As brittle materials, glasses are also quite susceptible to failure due to thermal shock.

Other mechanical properties of glasses are inherent to the material. The elastic modulus, E, is determined by the individual bonds in the material and by the structure of the network. The hardness of glasses is a function of the strength of individual bonds and the density of packing of the atoms in the structure.

ELASTIC MODULUS

As classic brittle materials, glasses exhibit nearly perfect Hookian behavior on application of a stress. The ratio of the strain, ε, resulting from application of a stress, σ, is a constant which is known as the *elastic modulus*, or *Young's modulus*, E, which is defined by the expression

$$\sigma = E\varepsilon \tag{9.1}$$

If a tensile stress is applied to a specimen in the direction of the x-axis, the specimen will elongate in that direction. This elongation will be accompanied by contraction in the y and z directions. The ratio of the transverse strain to the axial strain is called *Poisson's ratio*. Poisson's ratio for oxide glasses generally lies between 0.2 and 0.3, although the value for vitreous silica is only 0.17. The *shear modulus*,

G, which relates shear strain, γ, to shear stress, τ, is given by the expression

$$\tau = G\gamma \tag{9.2}$$

Young's modulus, the shear modulus, and Poisson's ratio are related by the expression

$$G = \frac{E}{2(1+\nu)} \tag{9.3}$$

where ν is Poisson's ratio for the material.

The elastic modulus of a material arises from the relation between an applied force and the resultant change in the average separation distance of the atoms which form the structure of that material. If we consider the Condon–Morse curve for force, F, as a function of atomic separation distance, r, we can write an expression of the form

$$F = -\frac{a}{r^n} + \frac{b}{r^m} \tag{9.4}$$

where a, b, n, and m are constants. The curve showing the variation of force with separation distance (Figure 9.1) passes through zero at a distance r_0, which is the equilibrium spacing of the atoms. Application of a macroscopic force will increase the average spacing by an amount

$$\varepsilon = \frac{(r - r_0)}{r_0} \tag{9.5}$$

where r is the strained interatomic distance. The elastic modulus must then be equal to the slope of the force–distance curve at r_0.

The simple model based on the Condon–Morse curve applies quite well to highly ionic, close packed structures. If we consider the structure of glasses, we find that the modulus is also influenced by the dimensionality and connectivity of the structure, with a trend toward increasing elastic moduli as the structure changes from a chain structure to a layered structure to a fully connected three-dimensional network. Weak bonds between chains or layers effectively offset the influence of the strong bonds between atoms within the building blocks of the structure and allow easier distortion of the structure. The presence of breaks in the linkage within a structure, *e.g.* non-bridging oxygens, also allows easier displacement of atoms and reduces the elastic modulus. Replacement of modifier ions by aluminum ions, which reduce the non-bridging oxygen concentration and increases the connectivity of the network, also increases the elastic modulus of silicate glasses. The highest elastic

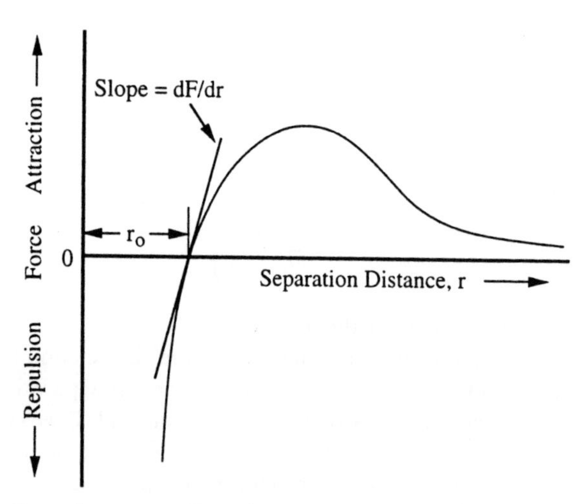

Figure 9.1 *Force–distance curve illustrating the origin of the elastic modulus*

moduli for oxide glasses are found in glasses such as the rare earth or yttrium aluminosilicates, which feature strong bonds and high packing densities. Nitriding of these glasses, which provides three-coordinated nitrogen linkages between tetrahedra, further increases the elastic modulus, with very high values found for glasses in SIALON (silicon aluminum oxynitride) systems. Values for the elastic modulus of inorganic glasses typically range from 10 to 200 GPa.

Since the elastic modulus of glasses is related to bond strength, it is not surprising to find that glasses with high glass transformation temperatures usually also have high moduli. Furthermore, if we recall the discussion of the thermal expansion coefficient in Chapter 7, where it was shown that the thermal expansion coefficient is also explained using a Condon–Morse diagram, it should not be too surprising to learn that low expansion glasses often have high elastic moduli.

HARDNESS

The hardness of glasses is usually defined in terms of either the scratch hardness using the *Moh's scale* or *indentation hardness* using a *Vickers* indenter. Oxide glasses lie in the range of 5–7 on Moh's scale, *i.e.*, they will scratch apatite (hardness of 5) but will not scratch crystalline quartz (hardness of 7). The Vickers hardness of oxide glasses ranges from 2 to 8 GPa, with values of over 11 GPa for nitrided glasses. These values are much lower than the Vickers hardness of diamond, which is $\approx$100 GPa.

Borate, germanate, and phosphate glasses are typically softer than silicate glasses. Chalcogenide glasses are much softer, with Vickers hardness values in the range of 0.3 GPa for vitreous selenium to just over 2.0 GPa for the three-dimensional structures found for Ge–As–S glasses. In general, the effects of glass composition on hardness parallel those found for elastic modulus.

FRACTURE STRENGTH

The fracture strengths of glasses are usually far less than their theoretical strengths. Fracture strength can only be described in terms of a distribution function and does not exhibit a single value characteristic of a given glass composition. The reduction in strength is attributed to surface flaws which severely weaken the glass.

Theoretical Strength of Glasses

The theoretical strength of a material is given by the force which must be applied to overcome the maximum restorative force predicted by Equation 9.4. Once the interatomic separation distance exceeds the distance corresponding to the maximum restorative force, continued application of force will extend the bond distance until the bond is broken and a crack can propagate through the material. Orowan proposed that the stress necessary to break a bond is determined by the energy necessary to create two new surfaces due to the fracture. The Orowan stress, σ_m, is given by the expression

$$\sigma_m = \sqrt{\frac{E\gamma}{r_0}} \tag{9.6}$$

where γ is the fracture surface energy, which has a value in the range 2–4 J m^{-2}. If we substitute values of $E = 70$ GPa, $\gamma = 3$ J m^{-2}, and $r_0 = 0.2$ nm into this expression, we obtain a theoretical strength of 32 GPa for a typical silicate glass. Since the terms in this expression are all relatively independent of glass composition, we thus predict that glasses should have strengths in the range of 1–100 GPa, regardless of composition.

Practical Strengths of Glasses

The strengths calculated using Equation 9.6 are orders of magnitude greater than those found in practical applications of bulk glasses. This reduction of strength is attributed to the presence of flaws in the surface of the glass. These flaws act as stress concentrators, increasing the local stresses to levels exceeding the theoretical strength and causing fracture of the glass. Griffith treated this problem in detail and derived the expression

$$\sigma_f = \sqrt{\frac{2E\gamma}{\pi c^*}} \tag{9.7}$$

where σ_f is the failure stress and c^* is the critical crack length for crack growth. Attainment of the critical crack length is only a necessary condition for crack growth. It is also necessary for the stress at the crack tip to exceed the theoretical strength of the material before the crack will grow spontaneously. Since Griffith flaws typically have curvatures approaching atomic dimensions at their tips, Orowan argues that any applied stress sufficient to exceed the Griffith criterion will also exceed the theoretical strength of the material and that the Griffith criterion is usually sufficient to cause fracture.

We have already argued that the elastic modulus and the fracture surface energy are relatively small functions of glass composition. Flaws, which are introduced by external factors, are not intrinsic to the material. Flaw lengths are determined by prior treatment of the surface and can vary over several orders of magnitude. It follows that the inherent strength of a glass is usually of little importance in determining the practical strength. The hardness of a glass can influence the practical strength through its influence on the resistance to flaw formation, *i.e.* scratch resistance.

Flaw Sources and Removal

How are the critical, or Griffith, flaws introduced into glasses? Obviously, contact with any material which is harder than the glass can cause a flaw. Abrasion with hard materials thus degrades the strength of a glass. Actually, contact with another piece of the same glass or with metal objects used to handle the glass is sufficient to generate flaws. Chemical attack can also generate flaws. Touching a glass with a fingertip will generate flaws through the attack on the surface due to the NaCl deposited from the skin. Thermal stresses induced during rapid cooling of a glass introduce flaws through thermal shock. Heating glasses for prolonged times has also been shown to reduce their strength by formation of a small number of surface crystals or by bonding of dust particles to the glass surface. In either case, a thermal expansion mismatch creates local flaws during cooling.

Protection of glass surfaces against flaw formation is very difficult. The surface of a freshly produced glass has a very high coefficient of friction for contact against other materials. Flaw generation can be reduced if a lubricant is applied to the fresh glass surface before any flaws are formed. Lubricating coatings are often applied to the surfaces of glass containers just after they exit from the annealing lehr. This coating must be resistant to wear, since any contact which penetrates the coating will result in flaw formation on the underlying glass.

Flaws can be removed by removing the outer surface of the material by chemical etching or mechanical polishing. Etching blunts the flaw tip and reduces the flaw length, while polishing simply reduces the length of the flaw to below the Griffith criterion. Flame polishing removes flaws through viscous flow in the near-surface region.

Strengthening of Glass

The strength of glasses can be increased by two methods. First, we can prevent the formation of flaws and remove those which do form. Removal of flaws is only effective for short times since new flaws are readily formed, while preventing their formation by use of coatings has proven to be of limited value. If we accept the fact that flaws will be present, we must concentrate on the prevention of crack growth. Since crack growth requires the presence of a tensile stress at the flaw tip, creation of a near-surface compression region should prevent crack growth. No growth will occur until the applied stress is large enough to overcome the residual compressive stress and produces a tensile stress at the crack tip.

Compressive surfaces can be produced by ion exchange (discussed in Chapter 8), *thermal tempering*, or by application or formation of a compressive coating. Thermal tempering involves the formation of a compressive layer by rapidly cooling a glass from at or above the upper limit of the glass transformation range. Since the interior of the glass will cool more slowly than the surface, the fictive temperature of the interior will be lower than that of the surface and the equilibrium density will be greater than that of the surface region. Since the interior and surface regions are bonded together, elastic strains must arise to counter the difference in equilibrium densities. The surface region is placed in compression, while the interior is placed in tension. The difference in fictive temperature is a function of the difference in cooling rate between the interior and surfaces of the glass, so that the magnitude of the compressive stress increases with increasing cooling rate and glass thickness. Consideration of the volume/temperature diagram for glasses reveals that the compressive stress also increases with the thermal expansion coefficient of the glass and with the difference between the thermal expansion coefficients of the glass and the supercooled liquid. It follows that thermal tempering is not very efficient for very thin wall containers or fibers because only a small difference in cooling rate occurs, or for low expansion glasses such as vitreous silica or many commercial borosilicate glasses, where the volume difference as a function of fictive temperature difference is small.

A compressive surface layer can be formed if a thin layer of material having a lower thermal expansion coefficient than the bulk glass can be created. Cooling the composite will create a compressive surface layer with a balancing tension region in the bulk glass. Application of a glaze can be carried out by fusing a thin sheet of a glass with a lower glass transformation temperature to the surface of a bulk glass or by more traditional glazing methods involving application of a low melting glass frit.

A variation of the ion exchange method using an ion which is smaller than that initially present in the glass can also produce a surface region of lower thermal expansion coefficient. If, for example, exchange of sodium from the glass with lithium from a bath occurs at temperatures above the T_g of the glass, relaxation will occur and chemical stuffing stresses will not exist. Since the surface region now consists of a glass containing lithium rather than sodium, the thermal expansion coefficient will usually be reduced in this region. Cooling the glass will force the lower expansion glass into compression, while the bulk glass is placed in tension.

The exchange of lithium for sodium offers another route to strengthening of glasses by formation of a surface crystallized region. If the glass is an alkali aluminosilicate, it may be possible to crystallize only the exchanged region, forming a very low thermal expansion coefficient phase such as virgilite or spodumene. Cooling the material will place the crystallized region in compression and the substrate glass in tension. Formation of a region which can be crystallized may also be possible after ion implantation of magnesium ions or by exchange of silver for sodium to form a region with a high nucleation density where crystallization will occur more readily than in the bulk glass.

Low expansion surface regions can be obtained by removal of alkali ions from the surface region of alkali–alkaline earth–silicate glasses. Exposure to SO_2 vapor, which is often carried out to improve chemical durability, leaches alkali ions from the surface, producing a silica-rich near-surface region. The reduction in alkali concentration reduces the thermal expansion coefficient and produces a compressive layer after cooling the glass.

Many other methods of strengthening are based on formation of composites by inclusion of fibers or whiskers or by crystallization to form glass-ceramics. Phase separation may also affect the strength by altering crack propagation mechanisms. Transformation toughening has also been attained by formation of a small concentration of zirconia crystals in glasses.

Statistical Nature of Fracture of Glass

Since the fracture strength of glass is usually controlled by the nature and concentration of the flaws present in the surface, it is not surprising to find a wide variation in measured strengths for a set of supposedly identical samples. Furthermore, since the propagation of a crack depends upon the simultaneous occurrence of a crack of sufficient length and a stress of sufficient magnitude, the experimental method used to measure strength affects the outcome of the measurement. Use of a three-point bend test, for example, yields more scatter in fracture strength data than does use of a four-point bend test. Consideration of the stress distribution in a rod used in these tests reveals that the maximum stress in a three-point bend test occurs at the point directly opposite the load point, while the maximum stress in a four-point bend test occurs over the region between the two load points. Since the area subjected to the maximum stress is much greater in the latter case, the probability of a critical flaw for a given stress occurring within the region of maximum stress is much greater.

Experimental results of failure stress studies can often be represented by a *Gaussian distribution*. According to Doremus, the probability, P, of finding a sample with a failure stress, S, is given by the expression

$$P = \left(\frac{1}{d\sqrt{2\pi}}\right)\exp\left[\frac{-(S - S_{\mathrm{m}})^2}{2d^2}\right] \tag{9.8}$$

where S_{m} is the strength of greatest probability and d is a measure of the width of the distribution (the standard deviation).

A second distribution function, called the *Weibull distribution*, is often used to describe fracture strengths. In this case, the fraction, F, of samples which fail at stresses below S is given by the expression

$$F = 1 - \exp\left[-\left(\frac{S}{S_0}\right)^m\right] \tag{9.9}$$

where S_0 is a scaling factor and m is a measure of the width of the distribution, frequently called the *Weibull modulus.* A small value of m represents a broad distribution. Doremus indicates that the Weibull distribution closely resembles a Gaussian distribution for values of m greater than ≈ 3.

For convenience in plotting data, the Weibull distribution expression given by Equation 9.9 is often converted to the form

$$\log[-\ln(1-F)] = m(\log S - \log S_0) \tag{9.10}$$

The data are then plotted as $\log[-\ln(1-F)]$ *versus* the load required for failure and the values of m and S_0 determined from a least-squares fit of the data. A plot of data in this form is called a *Weibull plot.* One should note that this expression may also be given as

$$\ln[-\ln(1-F)] = m(\ln S - \ln S_0) \tag{9.11}$$

FATIGUE OF GLASSES

The strength of glasses usually decreases with time under normal ambient conditions. This effect, known as *static fatigue*, is due to interaction of the glass with the surrounding atmosphere, resulting in crack growth under constant load. One also finds that a higher failure strength is observed when the load is increased rapidly than when it is increased slowly. Since this effect is observed under conditions of changing load, it is often called *dynamic fatigue.*

Both static and dynamic fatigue disappear for samples tested at liquid nitrogen temperatures. Since fatigue effectively disappears below $-100\,°C$, the use of liquid nitrogen simply provides a convenient method for obtaining very low temperatures and is not of particular relevance in fatigue studies. At higher temperatures, the time to failure for a given set of conditions decreases as the temperature increases. When tests are carried out at normal room temperatures, the rate of fatigue increases with inc.easing humidity.

Fatigue of silicate glasses is generally attributed to the *stress-enhanced reaction* of water with the silicate lattice at the crack tip, as expressed by the reaction

$$\text{Si–O–Si} + H_2O \rightarrow 2\text{SiOH} \tag{9.12}$$

This reaction between the silicate network and water molecules results in sharpening of the crack tip instead of lengthening of the crack. Since the reaction rate is essentially zero at very low temperatures, no fatigue occurs for testing at liquid nitrogen temperature (−196 °C). The increase in fatigue rate at higher temperatures is consistent with the increase in reaction rate expected with increasing temperature. Increases in humidity increase the fatigue rate by providing a higher concentration of reactant. Dynamic fatigue results reflect the requirement of sufficient time for the chemical reaction. If the load rate is increased rapidly, a higher stress will be reached before sufficient chemical reaction occurs to cause failure.

The simple model offered here explains the gross fatigue behavior of glasses, but does not explain some of the details of the process. Several other, more complex models have been offered to explain fatigue of glasses. A model proposed by Michalske and Freiman addresses the actual mechanism of the chemical reaction. Their model successfully predicts static fatigue in the presence of other molecules such as ammonia, while simultaneously explaining why fatigue does not occur in the presence of N_2 or CO. Another proposed mechanism known as the *chemical wedge* suggests that the reacting molecules do not actually reach the crack tip. Molecules entering the crack are drawn toward the tip by capillary action. The wedging action of these molecules increases the stress at the crack tip, causing rupture of the Si–O–Si bonds.

THERMAL SHOCK

Thermal shock is a serious problem wherever glasses are rapidly cooled over extended temperature ranges. A cooling rate gradient can lead to thermal tempering of glasses by producing different fictive temperatures in the surface and bulk of the glass. Unfortunately, cooling with a temperature gradient in a glass also produces temporary stresses which counter the permanent stresses due to differences in fictive temperature. If we consider a glass plate held at the glass transformation temperature, no stress will exist after some finite relaxation time and the fictive temperature will be T_g. If we were able to cool the surface of this plate instantaneously to room temperature, the volume in the surface region should shrink due to thermal contraction (the negative of thermal expansion during heating) to the room temperature value appropriate for a fictive temperature of T_g. If the center of the plate is still at the glass transformation temperature, however, the local volume will be considerably greater than that of the surface. After thermal equilibration at room temperature, the volumes of the surface and bulk should be approximately equal since their fictive temperatures are nearly equal because little change in fictive temperature will occur for a moderately slow cooling from T_g. (Remember, in tempering, we cool from a temperature well above T_g, so that regions of different fictive temperatures can be formed.) The sample should now be relatively free of stress, *i.e.*, any stresses occurring during cooling are temporary.

Although the stresses formed during cooling are temporary, failure can occurs due to the high stress which occurs when the surface and bulk temperatures differ. The maximum possible stress will be generated if the surface is instantaneously cooled, while the bulk is still at the original temperature. Under these conditions, the stress is given by the simple expression

$$\sigma = \frac{E\alpha\Delta T}{(1-\nu)} \tag{9.13}$$

where ΔT is the difference between the surface and bulk temperatures and α is the thermal expansion coefficient of the material.

In practical situations, one is usually more interested in the maximum possible ΔT which can exist without failure of the glass. By rearranging Equation 9.13, we can write the expression

$$\Delta T = \frac{\sigma_f(1-\nu)}{E\alpha} \tag{9.14}$$

where σ_f represents the stress necessary to cause failure and ΔT is the maximum temperature differential which can exist without failure of the sample.

Examination of Equation 9.13 reveals that the temporary stress which occurs during rapid cooling for a given temperature differential increases with increasing thermal expansion coefficient and elastic modulus. The best thermal shock resistence is thus found for low-expansion, low modulus glasses. The maximum temperature differential which can be used without sample failure will be very high for a very low thermal expansion glass such as vitreous silica. On the other hand, even high thermal expansion glasses such as vitreous boric oxide may not be susceptible to thermal shock failure if their glass transformation temperatures are very low. (As a first approximation, we can assume that no stress will occur above T_g since the relaxation time will be very short at higher temperatures. This assumption becomes less valid as the cooling rate increases.)

In reality, an instantaneous cooling rate cannot be obtained for a sample of finite size. If we consider the case of a plate cooled at a constant rate, ϕ, in K sec^{-1}, we will still generate a parabolic thermal gradient through the thickness of the plate. If the material has a thermal diffusivity given by $k/\rho C_p$, where k is the thermal conductivity of the material, ρ is the density, and C_p is the heat capacity at constant pressure, the stress at the surface is given by the expression

$$\sigma = \frac{E\alpha}{(1-\nu)}\left(\frac{\phi L^2 \rho C_p}{3k}\right) \tag{9.15}$$

where L is one half the thickness of the plate. Solutions to Equations 9.13 and 9.15 for a number of other geometries can be found in the literature.

ANNEALING OF THERMAL STRESSES

Our discussion of thermal stresses thus far has included both *temporary stresses* during cooling and *permanent stresses* deliberately introduced by thermal tempering. While the permanent stresses deliberately introduced by tempering are both desirable and carefully controlled, uncontrolled permanent stresses can be introduced by the same mechanisms during the formation of glasses. Not only can these stresses lead to delayed failure due to static fatigue, but they can also lead to dimensional changes due to relaxation during subsequent processing, stress birefringence, and refractive index gradients, which are undesirable for optical glasses. Since these stresses are accompanied by fictive temperature variations, other properties of glasses will also vary throughout the bulk of a specimen which has permanent stresses. Removal of these stresses, termed *annealing*, is thus essential for successful use of glasses in many applications.

Three problems must be addressed by any annealing schedule. First, the existing permanent stresses must be removed. Second, new residual stresses created during cooling through an annealing cycle must be minimized. Finally, temporary stresses which can cause failure during the annealing process must be avoided at all cost. These problems are usually addressed by heating the specimen to a uniform temperature in the glass transformation range, holding the specimen at constant temperature until the desired level of stress relief is attained, and then cooling the specimen at a rate which is sufficiently slow to avoid creation of new permanent stresses or fracture due to temporary stresses. The exact temperatures and heating and cooling rates used in creating such an annealing schedule are determined by the properties of the glass, its prior thermal history, and by the dimensions and shape of the specimen or product.

In general, the heating rate during the initial portion of the annealing schedule is relatively unimportant. Temporary stresses arising from a temperature gradient where the surface temperature exceeds the bulk temperature are compressive, so failure rarely occurs during heating of glasses. Glasses are typically held for 30–60 minutes at a temperature of 5 K above the annealing point obtained from viscosity measurements (see Chapter 6 for further discussion of annealing and strain points of glasses). If viscosity data are not available, a temperature of 5 K above the T_g obtained from a DSC or DTA measurement is usually adequate for annealing. After completion of the isothermal treatment, the glass is cooled very slowly, *e.g.* 1 K min^{-1}, through the glass transformation range to a temperature below the strain point. The rate of cooling is determined by the allowable final permanent stresses and property variations through the glass. *Fine annealing* of optical glass requires a slower cooling rate than that needed to produce the *coarse annealing* required for containers. Once the temperature is well below the strain point, changes in the fictive temperature are extremely slow, so that no additional permanent stresses are introduced by more rapid cooling. Subsequent cooling to room temperature can occur at any rate (10 K min^{-1} is common) below that where temporary stresses would cause failure.

SUMMARY

Glasses are brittle solids. The hardness of glasses is determined by the strength of the bonds forming the network and by the structure of the glass. Fracture of glasses is controlled by the presence of flaws and does not represent the inherent strength of the bonds between atoms. Glasses must be handled with care to maintain their strength, which may degrade by interaction with atmospheric gases. Thermal stresses may be either permanent or temporary and frequently cause fracture of glasses. Annealing of glasses is essential if reproducible products are to be created.

Chapter 10

Optical Properties

INTRODUCTION

Glasses are among the few solids which transmit light in the visible region of the spectrum. Glasses provide light in our homes through windows and electric lamps. They provide the basic elements of virtually all optical instruments. The worldwide telecommunication system is based on the transmission of light *via* optical waveguides. The esthetic appeal of fine glassware and crystal chandeliers stems from the high refractive index and birefringence provided by lead oxide, while the magnificent windows of many cathedrals exist only because of the brilliant colors which can be obtained in glasses.

The optical properties of glasses can be subdivided into three categories. First, many applications of glasses are based on bulk optical properties such as refractive index and optical dispersion. Other properties, including color, are based on optical effects which are strong functions of wavelength. Finally, modern glass technology increasingly relies on the application of non-traditional optical effects such as photosensitivity, photochromism, light scattering, Faraday rotation, and a host of others.

BULK OPTICAL PROPERTIES

The history of optical science closely parallels the history of the development of optical glasses. Development of early telescopes and microscopes immediately forced a search for new optical glasses with appropriate *refractive index* and *optical dispersion* characteristics. It can be argued that the development of modern astronomy, biology, and medical science were controlled by the ability of glass makers to develop glasses with the appropriate optical properties.

Refractive Index

The refractive index remains the most measured optical property of glasses as well as the most basic optical property for determination of the appropriate glass for many applications. The refractive index of any material is defined as the ratio of the velocity of light in a vacuum to the velocity of light in a medium. This ratio can be measured by application of Snell's law, which states that the refractive index, n, is given by the expression

$$n = \frac{\sin\Theta_i}{\sin\Theta_r} \tag{10.1}$$

where Θ_i is the angle of incidence and Θ_r is the angle of refraction for a beam of light striking the surface of a material. The refractive index can also be measured using methods based on the reflectivity of a surface, measurement of the critical angle for total reflection or Brewster's angle, or the Becke line technique. Further discussion of these methods can be found in many texts on optical properties.

The refractive index is not actually a constant, but varies with the wavelength of the incident light. The most commonly quoted index is usually designated as n_D and represents the index at the yellow emission line of sodium (589.3 nm). The index at the yellow emission line of helium (587.6 nm), designated n_d, is also commonly used. Since these wavelengths are nearly identical, there is very little difference between these indices.

The refractive index of glasses is determined by the interaction of light with the electrons of the constituent atoms of the glass. Increases in either *electron density* or *polarizability* of the ions increases the refractive index. As a result, low indices are found for glasses containing only low atomic number ions, which have both low electron densities and low polarizabilities. Glasses based on BeF_2 have refractive indices in the range of 1.27, while vitreous silica and vitreous boric oxide have refractive indices of about 1.458. At the other extreme, glasses with high lead, bismuth, or thallium contents may have refractive indices ranging from 2.0 to 2.5.

Since a majority of the ions in any glass are usually anions, the contribution to the refractive index from the anions is very important. Replacement of fluorine by more polarizable oxygen ions, or by other halides, increases the refractive index. Conversely, partial replacement of oxygen in oxide glasses by fluorine to form fluoroborate glasses, for example, reduces the refractive index. Since non-bridging oxygens are

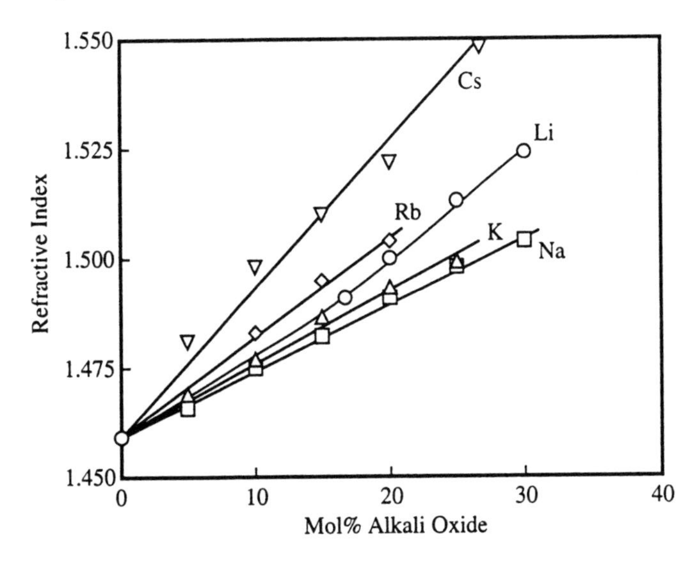

Figure 10.1 *Effect of composition on the refractive index of alkali silicate glasses*

more polarizable than bridging oxygens, compositional changes which result in the formation of non-bridging oxygens increase the refractive index of glasses, while changes in composition which reduce the non-bridging oxygen concentration can reduce the refractive index. The refractive indices of alkali silicate glasses (Figure 10.1) thus increase with increasing alkali oxide concentration, while replacement of alkali oxides by alumina, which reduces the non-bridging oxygen concentration, can cause a reduction in the refractive index.

The polarizability of the cation present increases as the field strength of the ion decreases, so that glasses containing cesium have a higher refractive index than those containing sodium. The most polarizable ions have very large electronic clouds and small oxidation numbers, *e.g.* Tl^+ and Pb^{2+}, which are used to produce very high refractive index glasses. Glasses which contain very high PbO concentrations, such as those found in unusual systems such as the $PbO–Ga_2O_3$ binary and the $PbO–Ga_2O_3–Bi_2O_3$ ternary, have refractive indices in excess of 2.5.

The density of a glass also plays a role in controlling the refractive index. Decreases in fictive temperature, which increase the density of most glasses, increase the refractive index. Since the fictive temperature is determined by the cooling rate through the glass transformation region, the refractive index is found to increase with decreasing cooling rate. This effect can be very important for optical applications, where fine annealing is essential to minimize local index variations. The refractive index also increases when glasses are either reversibly or irreversibly compacted by pressure or by exposure to high-energy radiation.

Thermal expansion of glasses can result in either an increase or a decrease in the refractive index. The density of a glass will decrease if it expands upon heating, which should decrease the refractive index. The polarizability of the ions, however, increases with temperature, which increases the refractive index and may therefore offset the effect of the decreasing density. Glasses with high thermal expansion coefficients and low temperature variations in polarizability are usually found in systems containing fluorine, such as the fluoride, fluorophosphate, or fluorosilicate systems. These glasses have negative coefficients for the variation of refractive index with temperature, $\mathrm{d}n/\mathrm{d}T$. Glasses with low thermal expansion coefficients and higher temperature variation of polarizability, as is the case for most silicate and borate glasses, have positive temperature coefficients of refractive index. These variations in refractive index are reversible so long as no relaxation of the density occurs during the temperature excursion.

Molar and Ionic Refractivities

The *molar refractivity* is directly proportional to the polarizabilities of the constituent ions of a glass. It can be shown that the molar refractivity, R_m, is given by the expression

$$R_m = V_m\left(\frac{n^2 - 1}{n^2 + 2}\right) \tag{10.2}$$

where V_m is the *molar volume* of the glass and n is the refractive index at the wavelength of measurement. The molar volume is equal to the molecular weight of the glass divided by its density.

The molar refractivity of a compound can be calculated from the contributions of each of the constituent ions. The molar refractivity for the compound A_xB_y for example, is given by the sum of the *ionic refractivities* of the constituent ions, R_n, times their concentration in the compound, or, in this case

$$R_m = xR_A + yR_B \tag{10.3}$$

Since the ionic refractivity depends on the polarizability of the ion, large values are found for the large, low field strength ions such as Tl^+ and Pb^{2+}. Variations in the ionic refractivity explain many of the major trends in the refractive index of glasses.

Although this method of estimating the molar refractivity works well for many inorganic compounds, it is difficult to apply to oxide glasses. The ionic refractivity of oxygen depends upon its role in the glass structure, so that the values for bridging and non-bridging oxygens are not identical. Furthermore, the ionic refractivity of oxygen ions depends on the nature of the associated cations. As a result, one can only use ionic refractivities as a guideline to the choice of ions for altering the refractive index of oxide glasses and not for quantitative calculations.

Since a typical glass contains from 50 to 80 atomic percent of anions, the ionic refractivities of the anions are very important in controlling the molar refractivity. The polarizabilities of the common anions increase in the order $F^- < OH^- < Cl^- < O^{2-} < S^{2-} < Se^{2-} < Te^{2-}$. This trend in ionic refractivities explains why replacement of oxygen by fluorine in fluoroborate, fluorosilicate, or fluorogermanate glasses decreases the refractive index of glasses, even though two fluorine ions are required to replace a single oxygen ion. The high refractive indices of chalcogenide glasses stem directly from the high ionic refractivities of sulfur, selenium, and tellurium ions.

The use of the molar refractivity stresses the role of ionic packing in controlling the refractive index of a glass. Since the refractive index is proportional to the molar refractivity divided by the molar volume, it is obvious that a small molar volume will yield a larger refractive index for a glass consisting of ions of similar polarizabilities. An example of this effect can be found in the refractive indices of many glasses containing lithium compared to similar glasses containing sodium or potassium (Figure 10.2). Since lithium actually causes a contraction of the vitreous network in many glasses, the molar volume is reduced by addition of lithium ions. This reduction in molar volume more than offsets the lower ionic refractivity of the lithium ion relative to that of sodium or potassium ions and results in a glass with a higher refractive index. In many cases, apparently anomalous trends in refractive index are resolved when the data are converted to molar refractivities.

Tables of optical data for glasses often include values for the specific refractivity, which is given by the expression

$$R_s = \frac{1}{\rho}\left(\frac{n^2 - 1}{n^2 + 2}\right) \tag{10.4}$$

where ρ is the density of the glass. The specific refractivity is primarily used for designing optical systems, where the mass of glass used may be important.

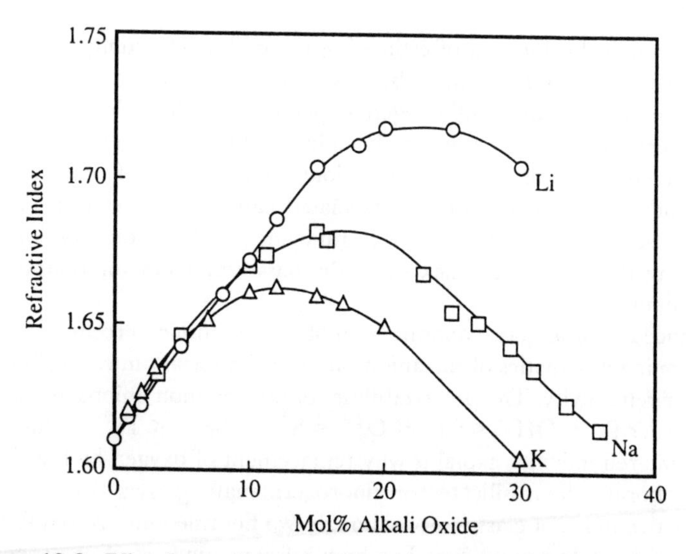

Figure 10.2 *Effect of composition on the refractive index of alkali germanate glasses*

Dispersion

The variation in index with wavelength, known as *optical dispersion* or simply as *dispersion*, is critical in the control of chromatic aberration of optical lenses. Ideally, dispersion is described by the entire curve of refractive index *versus* wavelength over the desired wavelength range. In general, however, it is more convenient to measure the refractive index at a few specified wavelengths and use these measurements as the basis for terms which can be used to compare the dispersion of different glasses. Using this approach, the mean dispersion is defined as the difference between the refractive indices measured at the F and C emission lines of hydrogen, which occur at 486.1 and 656.3 nm, respectively, or

$$\text{mean dispersion} = n_F - n_C \tag{10.5}$$

where n_F and n_C are the refractive indices for the indicated emission lines. The Abbe number, ν, which is commonly used to compare dispersions, is defined in various sources as either

$$\nu = \frac{n_D - 1}{n_F - n_C} \tag{10.6}$$

or

$$\nu = \frac{n_d - 1}{n_F - n_C} \tag{10.7}$$

These values are nearly identical.

More detailed information regarding the dispersion curve as a function of wavelength is often provided in the form of an expression of the form

$$n = C_0 + C_1\lambda^2 + C_2\lambda^{-2} + C_3\lambda^{-4} + C_4\lambda^{-6} \tag{10.8}$$

where C_0, C_1, C_2, C_3, and C_4 are constants. This expression is quite useful for many glasses, but should not be used in regions near any absorption bands.

ULTRAVIOLET ABSORPTION

Even transparent, colorless glasses cannot transmit radiation at wavelengths beyond their inherent *ultraviolet edge.* This frequency is believed to be due to the transition of a valence electron of a network anion to an excited state. Conversion of a network anion from the bridging state to a non-bridging state will lower the energy required for the electronic excitation and shift the ultraviolet edge to lower frequencies. The addition of alkali oxides to silica, therefore, results in a shift of the ultraviolet edge toward the visible region of the spectrum. Since initial additions of alkali oxides to boric oxide result in conversion of boron from three- to four-fold coordination, thus strengthening the network bonds, the ultraviolet edge does not shift toward the visible. Once the concentration of alkali oxide becomes sufficient to produce non-bridging oxygens, the expected shift of the edge toward the visible with increasing alkali oxide content is observed.

The ultraviolet edge of vitreous germania is closer to the edge of the visible spectral region than that of the other common oxide glassformers. Addition of large concentrations of alkali oxides shifts this edge to a frequency very near the visible. If these glasses are heated, they gradually become yellow with an increase in the intensity of the color with increasing temperature. The glasses return to the colorless state on cooling. This effect, known as reversible *thermochromism,* is due to the shift of the ultraviolet edge into the visible region at elevated temperatures.

In reality, the inherent ultraviolet edge of a glass is rarely observed. Very small concentrations of iron and other impurities result in very intense absorption bands. Since the absorption of energy is due to the transfer of an electron from the cation to a neighboring anion, these absorptions are said to be due to a *charge transfer transition* and the absorption band is called a *charge transfer band.* These bands are so intense that only their tail can be detected, so that the appearance of the spectrum is identical to that due to an inherent ultraviolet absorption. The impurity iron content of most silica used in glassmaking is so great that the inherent ultraviolet edge of silicate glasses is usually undetectable.

VISIBLE ABSORPTION

Absorption in the visible is perceived as color. A number of mechanisms exist for the creation of color in glasses. The most important commercial colored glasses contain either 3d transition metal ions or 4f rare earth (lanthanide) ions, where the coloration arises from the so-called ligand field effect. Other sources of color include the formation of metal or semi-conductor colloidal particles, optical defects induced by solarization or radiation, and charge transfer bands in the visible region of the spectrum.

Ligand Field Coloration of Glasses

Coloration of glasses by 3d transition metals ions is due to electronic transitions between normally degenerate energy levels of d-electrons. Since a detailed description of the mechanism leading to these electronic transitions (called *ligand field* or *crystal field* theory) can be found in many places, only a brief qualitative discussion will be provide here.

The 3d electronic levels are identical in energy for free ions. However, when a transition metal ion is surrounded by a few anions, called *ligands*, as in a crystal or glass, the interaction of the electric fields causes a small splitting of the energy levels. The magnitude of this splitting is a function of the field strength, number, and geometric arrangement of the neighboring anions. The number of different levels formed is a function of the electronic configuration and coordination number of the cation. Since the energy differences which commonly result for 3d transition metal ions from ligand fields are in the range of 1–3 eV, the absorption of photons by electronic transitions between split 3d levels results in visible coloration.

Similar arguments apply to the 4f electronic levels of the rare earth ions, where splitting of the 4f levels also produces absorption bands in the visible. Differences in the nature of the 3d and 4f ions result in less intense absorptions for the rare earth ions, as well as more complex spectra, which are due to the greater number of possible configurations of the seven 4f levels compared to the five 3d levels of the transition metal ions.

All of these electronic transitions are technically forbidden by Laporte's rule, which states that electronic transitions can only occur if the orbital angular momentum changes by ± 1 during the transition. Since this does not occur for transitions from one d state to another d state, or from one f state to another f state, no absorption should occur for these ions. Fortunately, Laporte's rule is relaxed in solids due to the lack of perfect spherical symmetry which results from the presence of a limited number of point sources, so that electronic transitions can occur with a low probability between 3d or 4f levels which are split by the fields of the neighboring ligands. The low probability of these transitions, however, does reduce the intensity of the absorption. As a result, ligand field induced transitions are much weaker than the charge transfer effects which occur in the ultraviolet.

Since the coloration of glasses by transition metal and rare earth ions results from ligand field effects, several general trends can be predicted. First, a change in oxidation state results in a change in the number of 3d or 4f electrons, resulting in a different number of possible electronic transitions for otherwise identical conditions. Since each possible electronic transition represents an absorption with a different energy, a difference in oxidation state will result in a different absorption spectrum.

Most 3d transition metal ions are found in either octahedral or tetrahedral coordination in oxide glasses. A change in coordination number will result in a difference in splitting energy and, depending upon the number of 3d electrons present, possibly a change in the number and relative positions of the potential electronic transitions.

Changes in the identity of the anions results in a change in their ligand field strength and thus a shift in the positions of the absorption bands with no change in their number or relative positions. The ligand field strength of the common anions decreases in the order $O^{2-} > F^- > Cl^- > Br^- > I^-$. In many cases, the transition metals appear to prefer to be associated with halide ions instead of oxygen ions in nominally oxide glasses. For example, the substitution of a small amount of NaCl for Na_2O in a sodium borate glass containing cobalt oxide can cause the color due to Co^{2+} ions to change from a dark blue–purple to lighter blue–green due to a small shift in the absorption band positions to longer wavelengths. Addition of a small amount of NaBr can result in a green glass, while additions of NaI can yield a red–brown glass. The Co^{2+} ions must preferentially associate with the small number of halide ions, since the color of the glass is actually due to a very small concentration of the transition metal ions.

The color is also altered by changes in concentration of the coloring cation, in the identity of the network former, and in the identity and concentration of the modifiers present. The effect of the concentration of the coloring ion is obvious: more *chromophores*, or coloring species, result in more absorption. The effects of changes in the network former and the modifier ions present are due to alterations in bond distance and bond strength between the coloring ions and the surrounding ligands. Replacement of a small diameter modifier ion by a larger one can also occasionally cause a change in the most favorable coordination number for the coloring ion.

Details of the coloration of glasses due to ligand field effects are further complicated by the possibility of redox interactions between two or more different transition metal ions. Other elements such as arsenic and antimony, which do not directly affect color, may alter the oxidation state of a coloring ion and alter the color of the glass. Changes in furnace atmosphere can also inadvertently alter the oxidation state of coloring ions due to changes in the concentrations of O_2, CO, CO_2, and H_2O vapor.

Amber Glass

Many glass containers have a brownish color popularly called 'beer-bottle brown'. This particular color occurs in glasses containing both iron and sulfur. Carbon is usually added to the batch to provide a reducing agent to insure the presence of sulfide ions. One model suggests that the coloration is due to an iron(III) ion in tetrahedral coordination with three O^{2-} and one S^{2-} ions. The actual absorption is due to a charge transfer process.

Control of amber browns in commercial glasses is quite difficult. The coloring agent, or chromophore, contains both an oxidized form of iron and a reduced form of sulfur. These forms can only co-exist in a melt in a narrow range of oxygen partial pressures. Since the intensity of the color will vary with oxygen partial pressure, reproducibility of the color is difficult. The oxygen partial pressure is usually controlled by varying the amount of carbon added to the melt or by controlling the redox of the combustion process. Replacement of sulfur by selenium changes the color from brown to black.

Colloidal Metal Colors

The red color produced in many glasses containing gold, known as *gold-ruby* glasses, is due to the presence of very fine colloidal gold particles.

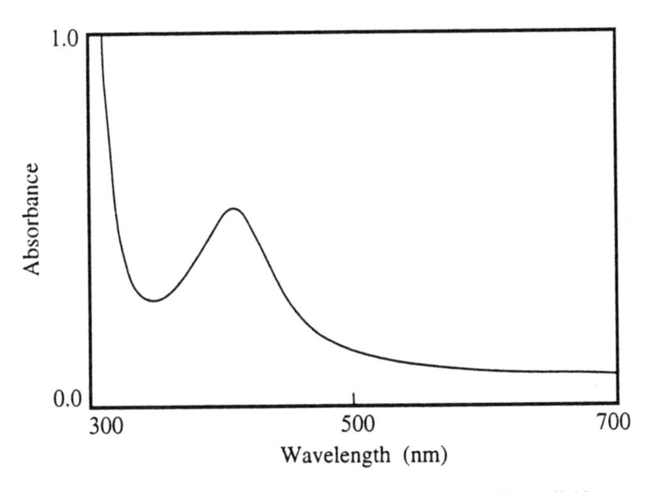

Figure 10.3 *Optical absorption spectrum for a glass containing silver colloids* (Data supplied by C. E. Lord)

The color is not due to light scattering, but rather to absorption by the particles, which cause an intense optical absorption band at about 530 nm. Doremus has calculated the shape and position of this band by assuming that the particles are spherical and using the optical properties of gold. He suggests that this band can be considered as a *plasma resonance* band, where the free electrons in the particles are treated as a bounded plasma. A similar absorption band, attributed to an identical mechanism, at 410 nm is obtained for glasses containing colloidal silver (Figure 10.3). The shift in band position results in a strong yellow coloration, which is called *silver-yellow* or *silver stain.*

A somewhat less esthetically pleasing red color can be produced in glasses containing copper. The absorption band due to copper occurs at 565 nm for these glasses and is similar in shape to those for gold and silver. While the red color of these glasses is usually attributed to copper colloids, others have proposed that the color is due to colloidal crystals of Cu_2O. Since both metallic copper and Cu_2O are often found in *copper-ruby* glasses, it is possible that the color arises from a combination of these species. Although the solubility of both gold and silver in silicate glasses limits the concentration of colloids which can be formed, the much higher solubility of copper permits the formation of a very large number of colloids. If the density of colloids is sufficiently high, the glass will be opaque rather than transparent.

A number of other colloidal species, including, but not limited to, Pb, As, Sb, Bi, Sn, and Ge, can be formed in glasses. The properties of the metals are such that these colloids result in brown, black, or gray colorations.

Colloids are usually formed by producing the glass with the metal in the ionic form and subsequently reducing the ions to form atoms. These atoms diffuse through the glass until they encounter other such atoms. The atoms then agglomerate to form nuclei, which grow to form the final colloids. Reduction can result from a redox reaction with other components of the glass or by reaction with an external reducing agent such as H_2. Many ruby glasses contain SnO_2, which provides an internal reducing agent. At the high temperatures used in melting, the oxidation equilibrium favors the production of Au^+ and Sn^{2+} ions. If the temperature is lowered, the equilibrium shifts toward the reduction of the gold and the oxidation of the tin, as shown by the reaction

$$Au^+ + Sn^{2+} \rightarrow Au^0 + Sn^{4+} \qquad (10.9)$$

This reaction, which is called *striking*, occurs spontaneously upon reheating an originally colorless glass to the correct temperature. A similar process can be used to produce glasses colored by silver or copper. The color is distributed uniformly throughout the glass.

Reduction by an external agent will occur if glasses containing gold, silver, or copper ions are exposed to H_2 gas at temperatures near the glass transformation range. Since reduction will occur in the near-surface region and grow into the glass, the color will occur in a layer at the glass surface. The thickness of this layer increases with the square root of time, indicating that hydrogen diffusion is important in controlling the coloration process. Although formation of colloids of other metals is difficult by use of an internal reducing agent, formation of a surface layer containing Pb, As, Sb, and Bi colloids is quite easy using an external reducing agent such as hydrogen gas.

Silver colloids can also be formed in the surface region of a glass by interdiffusion of silver from an external source with sodium or other alkali ions in the glass. The silver can be supplied from either metallic silver films or from molten silver salts. Since the exchange process requires that the silver be present as ions, metallic films must be heated in air or other sources of oxygen to temperatures above $\approx 150\,°C$ to form Ag_2O. Use of metallic films allows the production of complex images in the surface of the glass by sputtering the film through a mask. After ion exchange is completed, the glass is exposed to hydrogen to reduce the silver and create the colloids.

Silver colloids will form spontaneously if silver films on float glass are heated in air to temperatures above 300 °C. The process involves ion exchange between the silver ions and sodium ions from the glass, followed immediately by reduction of the silver by tin(II) ions in the glass surface. The tin(II) ions are present due to diffusion into the glass from the molten tin bath used in producing float glass. This reaction is highly specific to the 'tin-surface' of float glass and only occurs within the outer few micrometers of the glass surface.

Colloidal Semiconductor Colors

A number of glasses ranging in color continuously from yellow to orange to red to black can be produced by doping the melt with various combinations of CdS, CdSe, and/or CdTe. Similar glasses are produced using a mixture of CdS and ZnS. The as-cast glasses are colorless and must be heat treated at ≈ 550–$700\,°C$ to 'strike' the color. The optical spectra of these glasses differ from those of the colloidal metal colored glasses, with a sharp cutoff of transmission in the visible or near infrared, instead of the absorption bands observed for glasses colored by gold, silver, or copper colloids. This cutoff in transmission is due to the formation of very small semiconducting crystals of various cadmium chalcogenides. The absorption of higher frequency light is due to absorption of all photons having energies greater than the band gap of the semiconductor. Since continuous solid solutions form, it is possible to adjust this band gap over a wide range of energies, giving rise to a variety of colors. It has also been shown that the color is dependent upon crystallite size, with a shift toward the red with increasing crystal radius.

Radiation-induced Colors

An extremely large number of optical defects can be formed in glass by exposure to high energy radiation. These defects consist of trapped electrons or holes either at pre-existing sites in the glass or at sites created by the bond-breaking action of the radiation. Most of these defects give rise to absorption bands in the ultraviolet region of the spectrum and hence do not cause visible coloration of the glass. In general, the optical absorption results from electronic states in the gap between the valence and conduction bands. Photons induce transitions between the valence band and the defect levels or from the defect levels to the conduction band. Since a number of defects are often simultaneously produced by the radiation, multiple, overlapping absorption bands usually occur, producing complex optical absorption spectra.

Although vitreous silica usually remains colorless following irradiation to very high doses, doped silicas can become colored through the formation of defects associated with impurities. Purple samples, for example, are formed if the glass contains a small amount of aluminum, due to the formation of *aluminum–oxygen hole centers (AlOHC)*. Other impurities, such as germanium or titanium, can also produce colored vitreous silica by formation of defect centers.

Most common silicate glasses become brown after irradiation. The color is due to formation of many defects, especially hole centers associated with the non-bridging oxygens present in glasses containing alkali or alkaline earth oxides.

These optical absorptions can be bleached, or thermally annealed, by heating to sufficiently high temperatures. The thermal stability of the defects differs widely, so that the elimination of one defect may occur at room temperature, while the elimination of another requires heating to near the glass transformation temperature of the glass.

Solarization

Coloration of glasses by exposure to sunlight is known as *solarization*. Although some of the defects produced by higher energy radiation can also be produced by ultraviolet radiation, the classic solarization of glasses is due to a radiation-induced change in the valence of manganese, *via* the reaction

$$Mn^{2+} + Fe^{3+} + \text{photon} \rightarrow Mn^{3+} + Fe^{2} \qquad (10.10)$$

Many years ago, manganese was frequently added to glasses to serve as a 'decolorizer' for iron-induced optical absorption. Since this practice is no longer common, modern glasses do not produce the deep purple color characteristic of Mn^{3+} ions after long-term exposure to sunlight. While less common, other pairs of ions, including Mn–As, Fe–As, and several couples involving cerium, can also produce optical absorption changes due to solarization. Solarization of modern glasses usually produces brown shades similar to those produced by higher energy irradiation.

INFRARED ABSORPTION

Absorption of light in the ultraviolet and visible regions of the spectrum is due to electronic transitions. While there are some lower energy electronic transitions in the infrared region of the spectrum, most optical absorptions in this region in glasses are due to vibrational transitions. These absorptions can be divided into three categories: impurity absorptions due to gases or bound hydrogen isotopes, the *infrared cutoff*, or *multiphonon edge*, and the fundamental structural vibrations.

The frequency, ν, of a vibrational absorption in a diatomic molecule is given by

$$\nu = \left(\frac{1}{2\pi}\right)\sqrt{\frac{F}{\mu}} \qquad (10.11)$$

where F is the force constant for the bond and μ is the reduced mass of the molecule, as given by the expression

$$\mu = \frac{m_1 m_2}{m_1 + m_2} \qquad (10.12)$$

where m_1 and m_2 are the masses of the two atoms forming the molecule. The force constant is proportional to the bond strength, while the reduced mass is determined by the atomic weights of the atoms present. This model predicts that a vibrational absorption will shift toward the infrared if the bond is weak or if the masses of the atoms are large. It follows that replacement of a small, highly charged, low atomic number atom by a large, low field strength, high atomic number atom will result in a significant shift toward the infrared. Replacement of hydrogen by deuterium, termed the *isotope effect*, which does not significantly alter the force constant, will shift the band toward the infrared due to the change in mass.

Infrared Absorption by Bound Hydrogen Species

Virtually all oxide glasses contain *hydroxyl* in various forms, while other molecular species may or may not be present. The primary absorption band due to Si–OH bonds occurs at 2730 μm for vitreous silica (Figure 10.4). Since this is a vibrational absorption, *overtones* occur at $\nu/2$, $\nu/3$, *etc.* Other bands due to hydroxyl arise from the combination of the Si-OH frequencies with fundamental Si–O vibrations. These overtone and combination bands are relatively weak and are not of much importance for thin samples. However, when one forms an extremely long (km) optical fiber from vitreous silica, these bands become significant and must be eliminated to reduce optical losses to levels acceptable for telecommunication systems. Many millions of dollars have been invested in the research leading to the effective elimination of these very weak infrared absorption bands.

Replacement of hydrogen by deuterium or tritium causes all of these

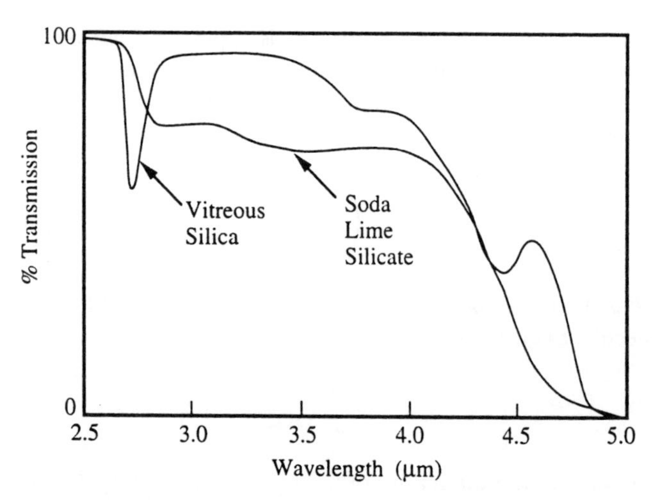

Figure 10.4 *Infrared transmission spectra showing bands due to hydroxyl in vitreous silica and soda–lime–silicate glasses*
(Data supplied by S. Chatlani)

bands to shift toward the infrared, as predicted by Equation 10.11. Replacement of Si^{4+} by B^{3+}, Ge^{4+}, Al^{3+}, or other ions also results in shifts of the band positions, with a larger shift due to germanium than due to the other elements.

Addition of alkali oxide to silica results in the formation of new bands due to hydroxyl as well as a shift in the position of the fundamental band toward the infrared. Hydroxyl bands are found at 2.75–2.95, 3.35–3.85, and 4.25 μm for common sodium silicate and soda–lime–silica glasses (Figure 10.4). The two bands at longer wavelengths are attributed to hydroxyl groups which are hydrogen bonded to neighboring non-bridging oxygens at two different distances. Replacement of alkali oxides by alumina, which eliminates the non-bridging oxygens from the structure, results in the elimination of the two bands attributed to hydrogen bonded hydroxyls. The hydroxyl spectra of alkali borosilicate glasses, which are often phase separated, with silica-rich and alkali borate-rich regions, are also very different from those of glasses containing large quantities of non-bridging oxygens.

Glasses can also contain bound hydrogen in the form of Si–H, B–H, and similar units. The fundamental vibration for Si–H occurs at 4.44 μm for vitreous silica. The absorption band is much sharper than those due to hydroxyl. These groups are usually found in glasses which have been melted under a hydrogen atmosphere, or which have been irradiated in the presence of H_2 gas. In both cases, the hydride groups can be removed by thermal treatment in air or vacuum at temperatures below the glass transformation range.

Hydroxyl can be formed in glasses by many methods. The most common form of hydroxyl, of course, stems from melting in the presence of water vapor and thus occurs for most commercial and laboratory melts. Formation of hydroxyl by reaction with water vapor can be described by the reaction

$$\text{Si–O–Si} + H_2O \rightarrow \text{Si–OH} + \text{HO–Si} \tag{10.13}$$

and by analogous reactions for other glassformers. Hydroxyl formed by reaction of water vapor with the melt is quite stable and can only be removed by long heat treatments in a water vapor free atmosphere at temperatures near and above the glass transformation range. Hydroxyl can also be formed by reaction with H_2, as described by the reaction

$$\text{Si–O–Si} + H_2 \rightarrow \text{Si–OH} + \text{H–Si} \tag{10.14}$$

The hydroxyl and hydride groups formed by these reactions are less stable than those formed by reaction with water molecules and can usually be removed at lower temperatures. The reaction described by Equation 10.14 can be driven either thermally during melting, or by irradiation at room temperature of glasses containing dissolved hydrogen. The thermal stability of the species formed is quite different, with a much lower temperature required for removal of the hydroxyl and hydride formed during irradiation.

Exposure of irradiated glasses to hydrogen gas after irradiation can also result in hydroxyl and hydride formation by reaction of H_2 molecules diffusing into the glass with radiation-induced defects. As a result, the defects are eliminated, the glass becomes colorless, and the infrared transmission is reduced. If the glass contains dissolved hydrogen during irradiation, no defects will be found after irradiation. This process, known as *chemical annealing*, can be used to eliminate optical defects in the ultraviolet and visible region for many glasses. Replacement of hydrogen with deuterium results in the formation of deuteroxyl instead of hydroxyl. In addition, the pre-existing hydroxyls in the glass will isotope exchange with the deuterium and become deuteroxyls.

Infrared Absorption by Dissolved Gases

Diatomic molecules containing only one element (H_2, O_2, N_2, *etc.*) do not absorb infrared radiation in the free gaseous state. It has been found, however, that hydrogen molecules dissolved in glasses cause a very weak infrared absorption band in silicate glasses in the region of 2.41 μm. This band, which is relatively symmetric and narrow for an infrared band in glasses, varies only slightly in position with glass composition for silicate glasses.

Dissolved carbon dioxide also causes an infrared absorption band in glasses. A narrow absorption band due to dissolved CO_2 molecules is found at 4.26 μm in sodium aluminosilicate and heavy metal fluoride glasses. Bands due to carbonate species formed by reaction of carbon dioxide with oxide melts have also been reported.

Infrared Cutoffs or the Multiphonon Edge

The infrared cutoff, or multiphonon edge, of glasses, is caused by the combinations and overtones of the fundamental infrared vibrations between the cations and anions which make up the glass structure. These extremely intense absorption bands prevent the practical application of glasses for transformation of light at longer wavelengths. The position of this edge is controlled by the strength of the bond between the atoms in the glass and the mass of those atoms (Figure 10.5). The edge wavelength shifts toward the infrared in the order $B_2O_3 < SiO_2 < GeO_2$ for the simple glassforming oxides. Traditional oxide glasses for infrared transmission are based on either germanate or calcium aluminate compositions, which transmit to ≈ 6 μm. Recently, the discovery of lead gallate and lead bismuth gallate glasses has extended the edge position for the best oxide glasses to ≈ 8 μm.

The elimination of oxygen, as in the heavy metal fluoride and chalcogenide glasses, permits the formation of glasses which transmit further into the infrared. Fluoride glasses typically have cutoff wavelengths in the range of 6–8 μm. Replacement of fluorine by chlorine, which both weakens the bonds and increases the average mass, extends this cutoff to 12–14 μm, while replacement by Br or I can shift the edge to ≥ 20 and ≥ 30 μm, respectively. Unfortunately, the glasses based on Br and I have such weak bonding that their physical and chemical properties are quite poor, preventing widespread application to date.

Chalcogenide glasses are frequently semiconductors, which means that they have a smaller band gap than those found for oxide glasses. In most cases, these glasses are actually opaque in the visible, with transmission only becoming measurable at >1 μm. The fundamental vibration frequencies for the network bonds are much lower than those found in oxide glasses, so the infrared cutoffs occur at much longer wavelengths. The cutoff wavelength increases in the order S < Se <

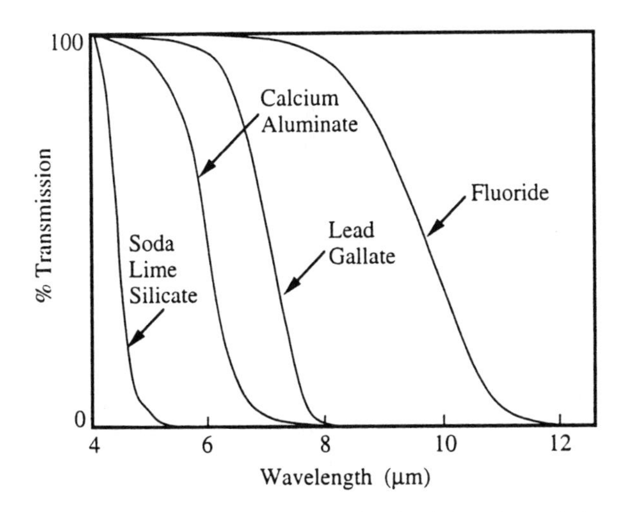

Figure 10.5 *Infrared transmission spectra showing the shift in the multiphonon edge for glasses of different compositions*

Te. Practical infrared windows which transmit to around 16 μm have been made from these glasses. Since all oxygen must be excluded during melting to form high quality materials, processing problems continue to limit the application of these glasses. The toxicities of Se and Te also increase the difficulty in processing commercial quantities of these glasses.

OTHER OPTICAL PROPERTIES OF GLASSES

Photosensitive and Photochromic Glasses

Photosensitivity and *photochromism* in glasses are closely related phenomena. *Photosensitive* glasses contain Au, Ag, or Cu as ions. When exposed to ultraviolet radiation, these ions are reduced to the atomic state. The reduction process is improved if the glass contains a small quantity of cerium oxide. Upon heating to a temperature where the atoms are mobile (usually between T_g and the softening temperature), agglomeration occurs as discussed for colloidally colored glasses. Since the colloids only form in the regions exposed to ultraviolet radiation, an image can be formed in the glass if a mask is used during the initial exposure to the radiation. Since the colloids do not redissolve into the glass, the image is permanent.

Photochemically machinable glasses can be produced from photosensitive glasses if the glass composition contains significant amounts of lithia and silica. After forming the image by precipitation and growth of silver colloids, further heat treatment can result in the growth of lithium metasilicate on the nuclei provided by the silver colloids. Since the nuclei only occur where the image existed, the crystals will form a positive copy of that image. Lithium metasilicate can be readily dissolved in mineral acids, at a rate which is considerably faster than that of the uncrystallized glass. As a result, leaching of the crystals occurs, producing holes through the glass in the form of a negative copy of the original image. The glass can be used in this state or reheated to a different temperature without formation of silver nuclei. If properly treated, lithium disilicate crystals will form, producing an opaque, strong glass-ceramic. This process is used for a variety of products, ranging from sophisticated electron beam masks containing thousands of holes per cm^2 to ordinary Christmas tree ornaments.

Photochromic glasses also contain silver, but, in addition, contain a halide. Small (8–15 nm) silver halide crystals are formed in the glass during the initial processing by heat treatment at temperatures above T_g. Exposure to ultraviolet light results in the formation of tiny anisotropic silver specks on the surfaces of the halide crystals. Since the halide is not removed from the vicinity of the silver, removal of the source of light allows the dissociation reaction to proceed in the reverse direction, which removes the darkening. Other photochromic glasses can be produced based on copper cadmium halides instead of silver halides. These glasses darken to a green color instead of the gray or brown obtained for the glasses containing silver. The reversibility of the darkening allows these glasses to be used for sunglasses which automatically darken outdoors, but clear on returning indoors.

Opal Glasses

Opal glasses are opaque due to light scattering resulting from either liquid–liquid or liquid–solid phase separation. Compositions in the lithium borosilicate system may spontaneously form opal glasses due to the coarse scale of liquid–liquid phase separation. Other opal glasses are formed by addition of a few percent of NaF or CaF_2 to the composition. Formation of many small crystals can occur spontaneously during cooling or during subsequent reheating of an initially clear glass. The opacity of opal glasses is controlled by the refractive index difference between the phases, the volume of the lesser phase, and the morphology of the composite. These glasses are used for decorative glass products such as vases and dinnerware.

Faraday Rotation

The terms *Faraday rotation* and the *magneto-optic effect* refer to the rotation of the plane of polarized light passing through a material when that material is exposed to a magnetic field. Faraday rotation can be used to rotate selectively the plane of polarization of light by varying an applied magnetic field. If a strong Faraday rotator is combined with a strong electromagnet and appropriate external polarizers, one can construct optical switches or other devices.

The angle of rotation of polarized light, Θ, passing through a sample of thickness, L, under the influence of a magnetic field of strength, H, is given by the expression

$$\Theta = VLH \tag{10.15}$$

where V is called the *Verdet constant.* This constant describes the strength of the Faraday effect for the particular material under study. A larger value of the Verdet constant will yield a larger rotation for a given magnetic field and sample thickness. The Verdet constant can be either positive or negative, which determines the direction of rotation. By convention, the Verdet constant is positive for a diamagnetic material and negative for a paramagnetic material.

The largest diamagnetic Verdet constants are found for glasses containing large concentrations of lead and bismuth in the systems $PbO–Ga_2O_3$ and $PbO–Bi_2O_3–Ga_2O_3$. The highest paramagnetic Verdet constants are found in glasses containing large concentrations of trivalent terbium, praseodymium, or dysprosium. Verdet constants in excess of -100 rad T^{-1} m^{-1} have been measured for terbium aluminosilicate glasses. These glasses have great potential for the production of superior rotator glasses due to their high rare earth concentrations, hardness, durability, and strength.

The Verdet coefficient of glasses varies linearly with concentration for mixtures of different rare earth ions. Since positive Verdet constants can be obtained for glasses containing Yb, Gd, or Sm, it is possible to produce glasses with a Verdet constant of exactly zero by the appropriate choice of composition.

SUMMARY

Many applications of glasses are based on the combination of a wide range of optical properties with ease of fabrication in simple and complex shapes. The optical properties of glasses influence their applications from the mundane desire for transparent containers for many products to the cutting edge of technology in the areas of telecommunications and the potential for optical computers. Although the discussion here is limited, the range of optical properties of glasses is so great that they must be considered among the leaders of high technology materials.

Chapter 11

Glass Technology

INTRODUCTION

Manufacturing of commercial glass products involves application of all of the principles discussed earlier in this book. Commercial production, however, requires expansion of these principles to very large scale melting and to the unique forming methods required for each type of product. Production of tubing from vitreous silica requires a completely different approach than production of bottles or window glass. Processing of traditional bulk consumer products differs greatly from the more sophisticated processing required for optical fibers and glass-ceramics. Non-melting methods of forming glasses, including vapor deposition and sol–gel processing, also deserve a brief discussion. Detailed discussions of glass technology can be found in several books dealing exclusively that subject.

CLASSICAL FORMING METHODS

Containers

Glass containers were originally hand blown. A *gob* of glass was gathered on the end of a *blowpipe* by dipping the pipe into a melt. Air was forced into the gob by blowing through the pipe, forming a hollow shell of glass around a bubble at the end of the pipe. The object was continually rotated and shaped by hand using paddles formed from wood which had been charred and soaked in water. Containers formed by this method were expensive and highly variable in quality.

The blowpipe method continues to be used to produce specialty items such as lead crystal goblets and other decorative glasses. Production of the vast quantity of containers used today, however, requires total automation of the same processes used by the glass blower. Mechanical gathering devices still form a gob on the end of a rod and transport the gob to a forming mold. The gob is cut free of the gathering rod and delivered to a split mold, where it is blown to rough shape. The object is transferred to a blow mold, where it is reheated and blown to the final shape. Variations on this process include initially pressing to form the rough shape instead of blowing for production of wide-mouth containers and the use of paste-mold processing, where the molds are dipped or sprayed with water before contacting the molten glass. After removal from the mold, the glasses are transferred to a continuous annealing lehr, where the stresses introduced during forming are relieved by relaxation.

Most common containers are made from soda–lime–silica glasses containing, in wt %, roughly 73% silica, 11% lime, 14% soda, and 2% alumina. These glasses also may contain very small quantities of MgO, K_2O, and SO_3, which is used as a fining agent. Specialized containers for chemicals and pharmaceuticals may be made from borosilicate glasses, which have an increased durability and do not release sodium into the contents through chemical attack on the glass.

Flat Glass

Flat glass was originally formed by hand by gathering a gob on the end of a pipe and then spinning the gob inside the furnace until it flattened due to centrifugal force (the *crown method*). Glass panes were cut from the outer portions of the disk. These panes were thinner at the outer edge of the disk and thicker toward the center of the disk. Installation of these panes with the thicker edge down has been proposed as an explanation for the observation that some windows in ancient cathedrals are thicker at the bottom edge. (The false attribution of this observation to flow of the glass remains one of the great myths of science.) The center of the disk, called the *bullseye*, was used as a crude lens or in decorative leaded glass windows.

Larger sheets of glass could be produced by blowing the gob into a cylindrical shape (*broad glass*) or by drawing a very large cylinder (*machine cylinder method*) from the melt. These cylinders were split parallel to their axis and flattened by reheating on a flat surface. The quality of the glass was poor, but larger sheets could be produced than by the crown method.

Continuous formation of flat sheets (*sheet glass*) can be performed by drawing a ribbon directly from the melt. Different commercial techniques involve variations in the method used to control the initial point of ribbon formation and in the direction of draw. All commercial processes involve initial vertical draws, but the ribbon is subsequently bent around a roller to a horizontal position in the L.O.F.–Colburn method. Difficulties in controlling the viscosity of the melt to the precision needed to produce consistent thickness ribbons can generate a 'waviness' in the glass. Modern improvements in process control have reduced this effect, so that excellent quality sheet glass can be obtained.

Plate glass was originally formed by casting molten glass on a metal table and rolling it into a sheet or by continuous rolling between water cooled rollers. Since the surface of the sheet reproduced the flaws of the rollers and table surfaces, the glass was ground and polished to a high degree of parallelism and to an optical finish. The resulting glass plates were optically superior to those produced by sheet methods.

Most modern glass sheet is produced using the float process (*float glass*), in which the melt flows from the glass tank directly onto the surface of a bath of molten tin. The upper surface of the melt is in contact with the atmosphere above the tin bath, which must be reducing to prevent oxidation of the tin. Since neither surface contains any flaws, the cooling melt forms a glass with an excellent surface finish. The glass has cooled sufficiently before leaving the bath that subsequent contact with rollers does not produce flaws others than minor scratches. The thickness of the glass is controlled by surface tension, with an equilibrium value of 7.1 mm. Glasses of other thicknesses are produced by either stretching or compressing the cooling melt. The final glass contains a significant concentration of tin in the surface exposed to the bath (*tin surface*), with a much smaller amount of tin in the surface exposed to the atmosphere (*air surface*). Much of the dissolved tin is in the tin(II) state, which gives rise to the ability of the tin surface of the glass to reduce silver ions introduced from an external source. The float process has been limited thus far to soda–lime–silica glasses.

The *fusion process*, or *downdraw method*, produces glass of similar quality to that produced by the float process. The melt flows from the tank into a shallow trough at a rate which causes the melt to overfill the trough and flow over each side. The melt rejoins just below the trough to form a single sheet of glass as it cools. The sheet thickness is controlled by the pull rate, so that quite thin (0.5 mm) sheets can be easily formed. While this process is slower than the float process, it is more versatile and has been used to produce glasses with a wide variety of compositions.

Patterned glasses are produced by rolling the cooling melt between a roller engraved with the negative of the desired pattern and a smooth roller. Glasses containing wire mesh are formed by a similar process, where the mesh in introduced into the molten glass just before it passes through the rollers.

Very thin (≤0.15 mm), or *microsheet* glass, such as that used for microscope cover slips, is produced by drawing the melt through a slit in a platinum plate. The glass thickness is controlled by the rate of pulling the cooled glass below the slit. This process is most suitable for very thin glass and is not used for thicker plates.

Glass Fibers

Glass fibers, or *fiberglass*, are widely used for thermal insulation, fire resistant textiles, and reinforcing fibers for composites. Chopped, or discontinuous, fibers can be produced by several techniques in which a thin stream of highly fluid melt flowing from the bottom of a container is broken into small segments by air jets, mechanical attenuation, or flame attenuation. While these methods are still occasionally used, the most commonly used method for production of discontinuous fiber, *rotary processing*, is similar to the process used to make 'cotton candy' from molten sugar. The glass stream falls into a spinning cylinder which has a large number of holes in its surface. Centrifugal forces extrude the melt through these holes into a high-velocity gas stream, which breaks the fibers into small lengths.

Continuous fibers are produced either directly from the original melt or from remelted glass marbles. The latter method allows production of the glass at a single site, with shipping of the relatively compact marbles instead of the much higher volume fiberglass. The melt is extruded through orifices in a platinum alloy '*bushing*', which are usually 0.8–3 mm in diameter, and then pulled to a fine diameter. These bushing contain hundreds of orifices, so many *filaments* are produced simultaneously. The filaments are pulled through a coating solution, called *sizing*, before gathering into a single *strand*.

A number of compositions are used to produce fiberglass. Insulation fibers are produced from modified soda–lime–silica compositions, which contain more alumina and iron than container glasses. Most of the continuous fiberglass is produced from so-called *E glass*, which actually includes a range of alkaline earth aluminoborosilicate compositions. This family of compositions was originally developed for applications requiring high electrical resistivity. If E glass fibers are exposed to strong mineral acids, they are leached until only a silica framework remains. If the leached glasses are consolidated by heat treatment, high-silica fibers are formed. Other common fiberglass products are made from *S* (for strong) *glass* and *C* (for corrosion-resistant) *glass*. S glass contains more alumina than E glass, while C glass contains more silica and less alumina than E glass.

Glass Tubing and Rod

Crude glass rods were originally drawn directly from melts by simply dipping the end of a metal rod into the melt and pulling upward. These rods were of poor quality, with variable diameters and cross-sectional shapes. The rods were often reworked by hand to improve their quality and stretched to form the desired diameter. Tubing was formed by using a blowpipe and forcing a bubble to travel down the rod as it was pulled from the melt.

Glass tubing is currently manufactured by flow of the melt through an orifice which is partially blocked by a bell-shaped blowpipe in the center of the orifice. Flow around the blowpipe produces a hollow rod whenever air is forced down the pipe. The diameter, wall thickness, and cross-sectional shape of the tubing, which need not be circular, are controlled by the design of the orifice/pipe combination and by the pulling speed applied to the tubing. Rods can be formed by simply stopping the flow of air through the pipe. Larger diameter and/or thick wall tubing is produced using an updraw process, where the tubing is pulled up instead of down, since the weight of the glass produces too much force on the tubing for large sizes.

Solid and Hollow Glass Spheres

Since glassforming melts are liquids, any droplet of melt allowed to fall freely through a sufficient distance will assume a spherical shape due to surface tension forces. Small spheres are routinely produced by allowing a stream of melt to flow through the bottom of a container. The stream is broken into small segments by an air jet or flame just below the container. If the temperature is high enough and the fall distance is great enough, these segments will become spherical before the melt solidifies. The same effect will occur if a pre-sized glass frit is used. In this case, the frit particles are heated in the upper portion of the vertical furnace, or *drop tower*, become fluid, and then are transformed to spheres as they fall through the hot zone. Frit particles can also be converted into spheres by blowing the frit through the flame of a gas jet.

Small glass spheres can also be formed by a variation of a process used to form glass fibers. A stream of melt is poured onto a rotating disk. The melt is thrown off the edge of the disk and broken into small segments. If the disk is cool and the surrounding temperature is low, the segments will remain in fiber form. If, however, the disk is heated and the surroundings are hot enough to allow the glass to form spheres before freezing to a glass, small spherical beads will be formed.

Large glass spheres (*marbles*) are formed by cutting small gobs from a melt stream. These gobs fall onto a pair of counter-rotating screws with thread depth equal to one half the desired marble diameter. The gobs are converted into spheres and cooled as the gobs travel down the length of the screw, where the finished marbles are collected.

Formation of hollow glass spheres requires the release of gas from the starting material during the sphere formation. The batch components are mixed in a liquid, which may include a '*blowing agent*' such as urea. This solution is spray dried to form uniform, but non-spherical, particles. These particles are then introduced directly into the flame from a burner or dropped down a vertical furnace. As the particles melt, the blowing agents decompose, releasing gases which blow the molten sphere into a hollow shell. As the melt begins to cool, it becomes impermeable to any remaining gases, which prevent the collapse of the shell.

Lamp Glass

Bulbs for electric lamps are produced using a *ribbon machine*. A stream of melt from the furnace passes through a set of rollers which flatten the stream into a ribbon. The rollers are machined such that heavier disks are formed at intervals along the ribbon. The moving ribbon is supported by a chain of plates which have holes at intervals matching the heavier disks on the ribbon. The glass from these disks sags through the holes, forming bubbles. Another chain carrying blowing orifices is brought into contact with the upper surface of the ribbon. Air puffs from these orifices expand the bubbles to a larger diameter. The ribbon now passes above a third chain carrying blow molds which close around the bubbles. Final puffs of air blow the bubbles into the lamp envelope shape. The mold is removed and the bulb is cracked from the ribbon with a hammer.

SPECIALIZED FORMING METHODS

Optical Fibers

Early optical fibers were produced using a *double-crucible* method, which closely resembles the method used to produce glass tubing. The outer, or *cladding*, glass would be supplied exactly as in tubing drawing. Suppose that the blowpipe used in tubing formation were used to supply a second melt of a different composition from the cladding glass inside the tube formed by the cladding melt. The result would be a glass rod consisting of the inner, or core, glass surrounded by the cladding glass. If the orifice is very small and the rod is pulled very rapidly, an optical fiber will form.

While some optical fibers are still produced using the double-crucible method, most are formed using a *preform* method. The most direct method for forming a preform consists of producing a rod of the core glass and a tube of the cladding glass. The diameter of the rod should be just slightly less than the inner diameter of the tube. After placing the rod inside the tube, the ensemble is heated to fuse the two glasses to form a unit.

Other solid preforms can be produced using casting methods. A cylindrical crucible is filled with the molten cladding glass. After partial solidification occurs from the crucible wall, the remaining melt is poured from the crucible. The core glass is now poured into the tube casting formed from the cladding glass. The crucible is removed to complete formation of the preform. A related process involves pouring only a small amount of the cladding melt in the crucible. The crucible is then rapidly rotated around the longitudinal axis so that the melt is forced against the crucible walls, forming a tube. After cooling, the core melt is poured into the cladding tube and the crucible removed to yield the preform.

Vitreous silica optical fibers for modern telecommunication systems must be made from ultrapure materials. Glasses produced from melts are incapable of reaching the quality levels required by these fibers. The glasses used in current fibers are produced *in situ* as the preform is formed by vapor deposition processes. Since the raw materials are liquids, purification by distillation can radically reduce impurity contents. Furthermore, since the glass never contacts crucible or refractory materials, the purity of the glass is maintained throughout the process.

Several variations on the basic vapor deposition process have been used to produce preforms. In the *OVD* (*outer vapor deposition*) method, the reaction of oxygen with metal halide vapors occurs while the materials are in the vapor state. Tiny *soot* particles (diameter of <200 nm) are formed by passing a stream of O_2 and $SiCl_4$ vapor through a flame. Dopants are added in the form of other metal halides, such as $GeCl_4$. A porous preform is gradually developed as the soot particles are deposited on the outside of a rotating vitreous silica rod. A refractive index gradient is formed by changing the concentration of dopants as the porous preform becomes larger in diameter.

After the porous preform is finished, the material is consolidated by sintering to form a solid rod. The target rod is removed and the blank is heated while Cl_2 gas flows through the center hole. The Cl_2 treatment is needed to remove residual water and hydroxyl by formation of HCl. After sintering, the final solid preform can be used to draw optical fibers.

The *MCVD* (*modified chemical vapor deposition*) method also relies on the production of glass from halide vapors. The deposition process occurs inside a vitreous silica tube, which is heated from the outside and which serves as the cladding for the fiber. The reaction of the vapors now occurs without contamination from gases from the flames, which never contact the deposited material. Consolidation of the soot occurs simultaneously with deposition. The process continues until the desired layer thickness is reached, after which the entire tube is collapsed by increasing the external temperature to complete the preform.

Once a preform is completed by any of the methods described here, or by one of the other numerous methods used for special materials, fibers are drawn by heating the preform in a furnace at the top of a fiber *drawing tower*. A starting fiber is drawn from the base of the preform. This starting fiber is fed into a mechanical drawing system, which includes automatic monitoring and feed-back control for maintaining uniform fiber diameter. The fibers are usually coated with a protective polymer film as they are drawn and before they contact any source of mechanical damage.

Glass-ceramics

A wide variety of materials have been produced using variations of the basic glass-ceramic process. The most common commercial glass-ceramics are based on lithium silicate, lithium aluminosilicate, magnesium aluminosilicate, sodium aluminosilicate, potassium fluoromica, and natural minerals and slags. Recent studies have dealt with the possible production of ceramic superconductors, ferrites, and optical materials using the glass-ceramic process. Glass-ceramics offer superior strength to glasses, can be produced with very fine and uniform grain sizes, and have zero porosity. Complex shapes can be made using traditional glassforming processes. The combination of multiphases allows adjustment of properties over a wide range of values.

By definition, glass-ceramics are first formed as glasses, using identical procedures to those used for other glasses. After the desired shape is formed, the product may either be cooled to room temperature for later reheating, or taken directly to a temperature where *nucleation* of crystals occurs at a well-characterized rate. Heterogeneous nucleation of the glass occurs at sites formed by an additive known as a *nucleating agent*, which may be a metal, titanate, zirconate, fluoride, or other species. The primary crystalline phase then nucleates on these heterogeneous nuclei. After nucleation, the object is heated to an optimum temperature for crystal growth and held until it reaches the desired degree of crystallinity. Subsequent surface treatments such as ion exchange or glazing with a lower-expansion glass may be carried out to increase the strength or provide decorative effects.

Glass-ceramics can also be formed from powdered glasses. Objects are formed by extrusion, pressing, spraying, or slip-casting and then heated to a temperature where nucleation occurs at the particle surfaces. These glass-ceramics undergo viscous sintering during the later stages of crystal growth to form pore-free materials. Glass-ceramic sealing frits can also be produced from surface nucleated glasses.

Porous Glasses

Porous glasses can be formed by sintering glass powders, by leaching of phase separated glasses, or by the sol–gel method. These glasses can be used in the porous state or can serve as precursors to fully consolidated glasses. Porous glasses are currently under intense investigation as potential selective separation membranes for a variety of gases and liquids. Impregnation of porous glasses before consolidation can be used to produce continuously graded glass seals, conductive glasses containing continuous carbon filaments, red glasses colored by colloidal spinel particles, and optical fiber preforms.

The formation of porous glasses by leaching of phase separated glasses is frequently called the Vycor® process, after the commercial material produced by Corning. These glasses are formed from phase separated borosilicate glasses which have microstructures consisting of two continuous phases. One of these phases is silica-rich, while the other contains most of the alkali and boron oxides. Since the alkali borate phase readily dissolves in hot acids such as HCl, HNO_3, or H_2SO_4, this phase can be leached from the glass by exposure to such acids, leaving a very porous silica-rich skeleton known as *thirsty glass*. The pores in thirsty glasses are often in the range of 2–10 nm in diameter and form continuous pathways through the glass. Internal surface areas may be as great as $200\ m^2\ g^{-1}$ of glass.

Dental Products

Glasses are used in a wide variety of dental products. Glass powders are used as fillers in composite resin materials. These glasses must have very fine particle sizes and typically contain strontium or barium to aid in *X*-ray photography of the completed restoration. Some dental cements also contain fine particles of glass powder. These glasses contain large concentrations of phosphates and fluorine.

Dental crowns and bridges consist of a metal substrate covered by multiple layers of porcelain and glass. The first layer of porcelain must be opaque to hide the metal, while subsequent layers provide color matching to the surrounding teeth and the correct luster and translucency to match the appearance of natural teeth. Good thermal compatibility between the various layers is essential to prevent thermal shock cracking and to minimize residual stresses which could lead to stress-corrosion failures or failure due to the combined residual and masticatory stresses.

Glass-ceramics are also used for dental prosthetics. These materials are primarily based on either lithium disilicate or calcium phosphate glasses. Machinable glass-ceramics based on fluoromica phases have also been developed for machined inlays. All of these materials must be stained to match the surrounding teeth to be acceptable to the patient.

Bioactive glasses are also under development for repair of bone. These glasses contain high concentrations of soda and lime and must have a high Ca/P ratio. These materials actually bond to the surrounding bone. While not in use as yet, these materials hold promise for a number of interesting dental applications.

Sealing and Solder Glasses

Many devices require the formation of seals between glasses and other glasses, metals, or ceramics. Compatibility in thermal expansion is usually the most basic criterion for selection of such glasses. Since the seal is formed at a high temperature, cooling through the transformation range to room temperature can produce high residual stresses unless good thermal contraction matches exist between the materials to be joined. These seals are called *matched seals*. It is also possible to make glass-to-metal seals which have large differences in thermal expansion coefficients between the glass and metal, provided the glass is placed in compression after cooling. These seals, called *compression seals*, are formed by placing a higher thermal expansion metal around a lower thermal expansion glass. The shrinkage of the metal during cooling thus places the glass in compression.

The softening temperature of the glass must also be compatible with the characteristics of the material to which it is joined. This requirement is somewhat difficult to meet for aluminum alloys, where the melting temperatures of the metals are relatively low for most glasses. Low-melting glasses based on phosphate systems have been developed for many of these alloys.

Solder glasses are specifically developed for high fluidity at low temperatures. They may be used as glasses (*vitreous solders*) or crystallized to form glass-ceramics (*crystallizing, or glass-ceramic, solders*) after seal formation. Solders are usually applied as frits which are heated to a temperature sufficient for good flow to form a uniform layer on the glasses to be joined. Vitreous solders have the advantages of short sealing times and the potential for salvaging of parts by remelting and reforming poor seals. Crystallizing solders are stronger and, after heat treatment, may be capable of use at temperatures near, or even higher than, the original sealing temperature.

Vitreous Silica Products

Vitreous silica is produced either from melting of natural quartz crystals (*natural silica* or *fused quartz*) or from the vapor phase reaction of oxygen with $SiCl_4$ (*synthetic silica*). The archaic term 'fused quartz' should be used with care, since those unaware of the jargon of the industry frequently misunderstand this term and believe that the glass is actually a crystalline material.

Commercial vitreous silicas are usually classed into one of four categories as a function of manufacturing method. Type I silicas are made from natural quartz by heating in an electric furnace. This material has a low hydroxyl concentration, but the impurity levels are typical of the starting material. Melting is carried out under hydrogen gas in a refractory metal container. The hydroxyl formed during melting is metastable and is removed by vacuum baking prior to sale of the material. Type II silicas are produced by flame heating of natural quartz, which produces more stable hydroxyl by reaction of water vapor formed in the flame with the silica network. The impurity levels are similar to those of type I silicas.

Synthetic silicas are classed as either type III (high hydroxyl) or type IV (low hydroxyl). Type III silicas are formed by combustion of $SiCl_4$ in an oxy–hydrogen flame, which contains a high concentration of water vapor. These silicas contain far more hydroxyl than any of the other four types, with concentrations of hydroxyl exceeding 1000 ppm by weight. Type IV silicas are produced by similar reactions, but with the use of dry flames, *e.g.*, oxygen plasmas, which reduce the hydroxyl content to less than 1 wt ppm. Since these materials are formed from highly purified liquids and gases, their impurity contents, other than hydroxyl and chlorine, are quite low.

Sol–Gel Processing

Sol–gel processing of glasses, especially vitreous silica, has been the subject of considerable research during the past two decades. Although hundreds of papers have been published in this field, practical applications are still limited due to the high cost and long times required to produce monolithic pieces of glass by these techniques. Although the details differ among the dozens of methods lumped under the term sol–gel, all sol–gel processes essentially involve formation of a sol containing the desired glass components, treatment of the sol in such a way that gelation occurs, and drying of the gel to remove all liquids. At this point, a very low-density, high-porosity solid exists which may be used for gas or liquid filtration or other applications. The dried gel is usually sintered at temperatures just above the glass transformation range to form a consolidated glass, which may still be slightly less dense than a glass of identical composition formed by melting.

Sol–gel processing allows the formation of glasses at temperatures well below those required for melting the same glass, reducing energy costs and increasing purity by reducing contact with refractories. The raw materials are high-purity liquids, so that the initial purity is usually much greater than that of solid batch components. Greater homogeneity should be possible due to initial mixing of liquids instead of powders. Low temperature gelation allows formation of near-final shapes, which can substantially reduce machining costs for the final product. Thin film application is easier than for glass frits and can be carried out at lower temperatures. On the other hand, the raw materials are often very expensive. Removal of the liquids from the pores is time-consuming and often results in cracking of the gel. Considerable shrinkage occurs during drying, which causes difficulties in forming complex shapes without further machining. The dried gels are also very weak and must be handled with great care.

At present, sol–gel technology appears best suited to applications involving either film or fiber formation. Formation of monolithic pieces of glass is usually more costly than by other methods. Monolithic glass products are limited to those which are expensive to produce anyway, such as vitreous silica, to complex shapes, where machining of glasses is expensive or very difficult, and to specialty compositions, which may not be formed using traditional melting processes.

SUMMARY

Many elements of glass processing are similar regardless of the product, so that an understanding of the basics of glass formation, melting phenomena, and transformation range behavior are essential to the production of almost all commercial glass products. Other basic phenomena, such as nucleation and crystallization and phase separation, serve as the basis for production of specialized products. Understanding of the relations between glass composition, thermal history, and handling and the properties of the glass is essential to the production of reproducible, high-quality products with the exact properties desired for any given application.

Bibliography

O. L. Anderson and D. A. Stuart, *J. Am. Ceram. Soc.*, 1954, **37**, 573.

C. A. Angell, *J. Non-Cryst. Solids*, 1988, **102**, 205.

R. J. Arujo, in 'Commercial Glasses', ed. D. C. Boyd and J. F. MacDowell, American Ceramic Society, Columbus, OH, 1986, ch. 12.

R. J. Arujo, in 'Treatise on Materials Science and Technology, vol. 12: Glass I', Academic Press, New York, 1977, ch. 2.

P. F. Aubourg and W. W. Wolf, in 'Commercial Glasses', ed. D. C. Boyd and J. F. MacDowell, American Ceramic Society, Columbus, OH, 1986, p. 51.

C. R. Bamford, 'Color Generation and Control in Glass', Elsevier, Amsterdam, 1977, ch. 1–11.

R. F. Bartholomew, in 'Engineered Materials Handbook, vol. 4: Ceramics and Glasses', ed. S. J. Schneider, Jr., ASM International, Materials Park, OH, 1991, p. 460.

W. C. Bauer and J. E. Bailey, in 'Engineered Materials Handbook, vol. 4: Ceramics and Glasses', ed. S. J. Schneider, Jr., ASM International, Materials Park, OH, 1991, pp. 378–385.

G. H. Beall, in 'Commercial Glasses', ed. D. C. Boyd and J. F. MacDowell, American Ceramic Society, Columbus, OH, 1986, p. 157.

P. P. Bihuniak, in 'Commercial Glasses', ed. D. C. Boyd and J. F. MacDowell, American Ceramic Society, Columbus, OH, 1986, p. 105.

Z. U. Borisova, 'Glassy Semiconductors', Plenum Press, New York, 1981, ch. 2–5.

N. F. Borrelli and T. P. Seward, in 'Engineered Materials Handbook, vol. 4: Ceramics and Glasses', ed. S. J. Schneider, Jr., ASM International, Materials Park, OH, 1991, p. 439.

S. A. Brawer, 'Relaxation in Viscous Liquids and Glasses', American Ceramic Society, Columbus, OH, 1985.

R. Bruckner, *J. Non-Cryst. Solids*, 1970, **5**, 123.

R. Bruckner, *J. Non-Cryst. Solids*, 1971, **5**, 281.

M. Cable, *J. Am. Ceram. Soc.*, 1966, **49**, 436.

J. W. Cahn and R. J. Charles, *Phys. Chem. Glasses*, 1965, **6**, 181.

J. Coon and J. E. Shelby, *J. Am. Ceram. Soc.*, 1988, **71**, 354.

D. E. Day and G. E. Rindone, *J. Am. Ceram. Soc.*, 1962, **45**, 489.

E. W. Deeg, in 'Commercial Glasses', ed. D. C. Boyd and J. F. MacDowell, American Ceramic Society, Columbus, OH, 1986, ch. 1.

M. R. DeSocio, 'Effect of Potassium Content on the Properties of Lithium Silicate Glass-Ceramics', MS Thesis, Alfred University, NY, 1985.

R. H. Doremus, 'Glass Science', Wiley, New York, 2nd edn., 1994, ch. 1–6, 8–10, 12–17.

P. J. Doyle, 'Glass-Making Today', Portcullis Press, 1979, ch. 1, 4, and 6.

W. H. Dumbaugh, in 'Engineered Materials Handbook, vol. 4: Ceramics and Glasses', ed. S. J. Schneider, Jr., ASM International, Materials Park, OH, 1991, p. 423.

W. H. Dumbaugh and P. S. Danielson, in 'Commercial Glasses', ed. D. C. Boyd and J. F. MacDowell, American Ceramic Society, Columbus, OH, 1986, p.115.

C. K. Edge, in 'Commercial Glasses', ed. D. C. Boyd and J. F. MacDowell, American Ceramic Society, Columbus, OH, 1986, p. 43.

T. H. Elmer, in 'Engineered Materials Handbook, vol. 4: Ceramics and Glasses', ed. S. J. Schneider, Jr., ASM International, Materials Park, OH, 1991, p. 427.

I. Fanderlik, 'Optical Properties of Glass', Elsevier, Amsterdam, 1983, ch. 1–7.

J. E. Flannery and D. R. Wexell, in 'Commercial Glasses', ed. D. C. Boyd and J. F. MacDowell, American Ceramic Society, Columbus, OH, 1986, ch. 11.

J. W. Fleming, in 'Experimental Techniques of Glass Science', ed. C. J. Simmons and O. H. El-Bayoumi, American Ceramic Society, Westerville, OH, 1993, ch. 1.

J. Francel, in 'Commercial Glasses', ed. D. C. Boyd and J. F. MacDowell, American Ceramic Society, Columbus, OH, 1986, p. 79.

E. J. Friebele, in 'Optical Properties of Glass', ed. D. R. Uhlmann and N. J. Kreidl, American Ceramic Society, Westerville, OH, 1991, ch. 7.

E. J. Friebele and D. L. Griscom, in 'Treatise on Materials Science and Technology, vol. 17: Glass II', Academic Press, New York, 1979, ch. 6.

G. H. Frischat, 'Ionic Diffusion in Oxide Glasses', Trans Tech Publications, Aedermannsdorf, Switzerland, 1975.

G. S. Fulcher, *J. Am. Ceram. Soc.*, 1925, **8**, 339 (reprinted in *J. Am. Ceram. Soc.*, 1992, **75**, 1043).

C. A. Gressler and J. E. Shelby, *J. Appl. Phys.*, 1989, **66**, 1127.

A. A. Griffith, *Phil. Trans. Royal Soc.*, 1921, **A221**, 163.

W. K. Haller, in 'Commercial Glasses', ed. D. C. Boyd and J. F. MacDowell, American Ceramic Society, Columbus, OH, 1986, p. 133.

W. Haller, D. H. Blackburn, and J. S. Simmons, *J. Am. Ceram. Soc.*, 1974, **57**, 120.

H. W. Hayden, W. G. Moffatt, and J. Wulff, 'The Structure and Properties of Materials, vol. III: Mechanical Behavior', Wiley, New York, 1965, ch. 2 and 7.

G. S. Henderson and M. E. Fleet, *J. Non-Cryst. Solids*, 1991, **134**, 259.

C. D. Hendricks, in 'Engineered Materials Handbook, vol. 4: Ceramics and Glasses', ed. S. J. Schneider, Jr., ASM International, Materials Park, OH, 1991, p. 418.

P. L. Higby, J. E. Shelby, and M. Suscavage, *J. Appl. Phys.*, 1985, **58**, 4142.

P. L. Higby, J. E. Shelby, J. C. Phillips, and A. D. LeGrand, *J. Non-Cryst. Solids*, 1988, **105**, 139.

M. Horton and J. E. Shelby, *Phys. Chem. Glasses*, 1993, **34**, 238.

M. D. Ingram, *Phys. Chem. Glasses*, 1987, **28**, 215.

H. Jain, in 'Experimental Techniques of Glass Science', ed. C. J. Simmons and O. H. El-Bayoumi, American Ceramic Society, Westerville, OH, 1993, ch. 12.

P. F. James, in 'Glasses and Glass-Ceramics', ed. M. H. Lewis, Chapman and Hall, London, 1989, ch. 3, p. 59.

P. F. James, in 'Nucleation and Crystallization in Glasses', ed. J. H. Simmons, D. R. Uhlmann, and G. H. Beall, American Ceramic Society, Columbus, OH, 1982, p. 1.

J. M. Jewell, J. Coon, and J. E. Shelby, *Mater. Sci. Forum*, 1988, **32–33**, 421.

J. M. Jewell, C. M. Shaw, and J. E. Shelby, *J. Non-Cryst. Solids*, 1993, **152**, 32.

W. D. Kingery, 'Introduction to Ceramics', Wiley, New York, 1960, ch. 15 and 18.

P. Klocek and G. H. Sigel, 'Infrared Fiber Optics', SPIE Optical Engineering Press, Bellingham, WA, 1989, sect. 2 and 3.

J. T. Kohli, in 'Rare Elements in Glasses', ed. J. E. Shelby, Trans Tech Publications, Aedermannsdorf, Switzerland, 1994, ch. 5.
J. T. Kohli, R. A. Condrate, and J. E. Shelby, *Phys. Chem. Glasses*, 1993, **34**, 81.
J. T. Kohli, J. E. Shelby, and J. S. Frye, *Phys. Chem. Glasses*, 1992, **33**, 73.
J. C. Lapp and J. E. Shelby, *Adv. Ceram. Mater.*, 1986, **1**, 174.
J. C. Lapp and J. E. Shelby, *J. Non-Cryst. Solids*, 1986, **86**, 350.
J. C. Lapp and J. E. Shelby, *J. Non-Cryst. Solids*, 1987, **95–96**, 889.
C. E. Lord, 'Crystallization and Properties of Lithium Aluminosilicate Glass-Ceramics', MS Thesis, Alfred University, NY, 1995.
J. Lucas and J. L. Adam, in 'Optical Properties of Glass', ed. D. R. Uhlmann and N. J. Kreidl, American Ceramic Society, Westerville, OH, 1991, ch. 2.
P. W. McMillan, 'Glass-Ceramic', Academic Press, London, 2nd edn., 1979, ch. 2, p. 7.
A. Margaryan and M. A. Piliavin, 'Germanate Glasses: Structure, Spectroscopy, and Properties', Artech House, Boston, 1993, ch. 3–6.
O. V. Mazurin and E. A. Porai-Koshits, in 'Phase Separation in Glass', ed. O. V. Mazurin and E. A. Porai-Koshits, North-Holland, Amsterdam, 1984, ch. 6.
O. V. Mazurin, G. P. Roskova, and E. A. Porai-Koshits, in 'Phase Separation in Glass', ed. O. V. Mazurin and E. A. Porai-Koshits, North-Holland, Amsterdam, 1984, ch. 4.
O. V. Mazurin, M. V. Strel'tsina, and T. P. Shvaiko-Shvaikoskaya, 'Properties of Glasses and Glass-Forming Melts', Izdatel'stvo Nauka, Leningrad, 1973, vol. 1.
O. V. Mazurin, M. V. Strel'tsina, and T. P. Shvaiko-Shvaikoskaya, 'Properties of Glasses and Glass-Forming Melts', Izdatel'stvo Nauka, Leningrad, 1975, vol. 2.
T. A. Michalske and S. W. Freiman, *J. Am. Ceram. Soc.*, 1983, **66**, 284.
R. E. Mould and R. D. Southwick, *J. Am. Ceram. Soc.*, 1959, **42**, 582.
C. T. Moynihan, *J. Am. Ceram. Soc.*, 1993, **76**, 1081.
K. Nassau, 'The Physics and Chemistry of Color', Wiley, New York, 1983, ch. 5 and 7–11.
NIST Certificate of Viscosity Values, Standard Sample No. 710, 1962.
NIST Certificate of Viscosity Values, Standard Sample No. 711, 1964.
NIST Certificate of Viscosity Values, Standard Sample No. 717, 1969.
E. Orowan, *Z. Krist.*, 1934, **A89**, 327.
A. Paul, 'Chemistry of Glasses', Chapman and Hall, London, 1982, ch. 3 and 7.
J. L. Piguet and J. E. Shelby, *Adv. Ceram. Mater*, 1986, **1**, 192.

L. R. Pinckney, in 'Engineered Materials Handbook, vol. 4: Ceramics and Glasses', ed. S. J. Schneider, Jr., ASM International, Materials Park, OH, 1991, p. 433.

P. R. Prud'homme van Reine, J. J. van den Hoek, and A. G. Jack, in 'Engineered Materials Handbook, vol. 4: Ceramics and Glasses', ed. S. J. Schneider, Jr., ASM International, Materials Park, OH, 1991, p. 1033.

H. Rawson, 'Inorganic Glass-Forming Systems', Academic Press, London, 1967, ch. 1–16.

H. Rawson, 'Properties and Application of Glass', Elsevier, Amsterdam, 1980, ch. 3–7.

G. E. Rindone, *Glass Ind.*, 1957, 561.

J. E. Ritter, in 'Engineered Materials Handbook, vol. 4: Ceramics and Glasses', ed. S. J. Schneider, Jr., ASM International, Materials Park, OH, 1991, p. 694.

R. M. Rose, L. A. Shepard, and J. Wulff, 'The Structure and Properties of Materials, vol. IV: Electronic Properties', Wiley, New York, 1966, ch. 13.

R. J. Ryder and J. P. Poole, in 'Commercial Glasses', ed. D. C. Boyd and J. F. MacDowell, American Ceramic Society, Columbus, OH, 1986, p. 35.

G. W. Scherer, 'Relaxation in Glass and Composites', Wiley, New York, 1986.

J. E. Shelby, *J. Am. Ceram. Soc.*, 1974, **57**, 436.

J. E. Shelby, *J. Non-Cryst. Solids*, 1979, **34**, 111.

J. E. Shelby, *J. Appl. Phys.*, 1979, **50**, 3702.

J. E. Shelby, *J. Appl. Phys.*, 1979, **50**, 8010.

J. E. Shelby, *J. Appl. Phys.*, 1980, **51**, 2561.

J. E. Shelby, *J. Appl. Phys.*, 1980, **51**, 2589.

J. E. Shelby, *J. Am. Ceram. Soc.*, 1983, **66**, 225.

J. E. Shelby, *J. Am. Ceram. Soc.*, 1985, **68**, 155.

J. E. Shelby, *J. Appl. Phys.*, 1986, **60**, 4325.

J. E. Shelby, in 'Experimental Techniques of Glass Science', ed. C. J. Simmons and O. H. El-Bayoumi, American Ceramic Society, Westerville, OH, 1993, ch. 10.

J. E. Shelby, *J. Non-Cryst. Solids*, 1994, **179**, 138.

J. E. Shelby, 'Rare Elements in Glasses', ed. J. E. Shelby, Trans Tech Publications, Aedermannsdorf, Switzerland, 1994, ch. 1, 2, 12, and 14.

J. E. Shelby, 'Handbook of Gas Diffusion in Solids and Melts', ASM International, Materials Park, OH, 1996, ch. 2 and 9.

J. E. Shelby and L. K. Downie, *Phys. Chem. Glasses*, 1989, **30**, 151.

J. E. Shelby and C. E. Lord, *J. Am. Ceram. Soc.*, 1990, **73**, 750.
J. E. Shelby and R. L. Ortolano, *Phys. Chem. Glasses*, 1990, **31**, 25.
J. E. Shelby and J. Vitko, *J. Non-Cryst. Solids*, 1982, **53**, 155.
J. E. Shelby, W. C. LaCourse, and A. G. Clare, in 'Engineered Materials Handbook, vol. 4: Ceramics and Glasses', ed. S. J. Schneider, Jr., ASM International, Materials Park, OH, 1991, p. 845.
J. E. Shelby, C. M. Shaw, and M. S. Spess, *J. Appl. Phys.*, 1989, **66**, 1149.
J. E. Shelby, J. Vitko, and C. G. Pantano, *Solar Energy Mater.*, 1980, **3**, 97.
R. D. Shoup, in 'Engineered Materials Handbook, vol. 4: Ceramics and Glasses', ed. S. J. Schneider, Jr., ASM International, Materials Park, OH, 1991, p. 445.
G. H. Sigel, in 'Treatise on Materials Science and Technology, vol. 12: Glass I', Academic Press, New York, 1977, ch. 1.
J. H. Simmons, in 'Experimental Techniques of Glass Science', ed. C. J. Simmons and O. H. El-Bayoumi, American Ceramic Society, Westerville, OH, 1993, ch. 11.
A. Smekal, *J. Soc. Glass Technol.*, 1951, **35**, 411T.
R. F. Speyer, 'Thermal Analysis of Materials', Marcel Dekker, New York, 1994, ch. 7 and 10.
A. N. Sreeram, A. K. Varshneya, and D. R. Swiler, *J. Non-Cryst. Solids*, 1991, **128**, 294.
J. E. Stanworth, 'Physical Properties of Glass', Oxford University Press, London, 1950, ch. 2 and 8.
V. R. Stenshorn, 'Effect of Hydroxyl Concentration on the Crystallization of a Lithium Silicate Glass-Ceramic', MS Thesis, Alfred University, NY, 1992.
H. J. Stevens, in 'Engineered Materials Handbook, vol. 4: Ceramics and Glasses', ed. S. J. Schneider, Jr., ASM International, Materials Park, OH, 1991, p. 394.
J. F. Stroman, in 'Engineered Materials Handbook, vol. 4: Ceramics and Glasses', ed. S. J. Schneider, Jr., ASM International, Materials Park, OH, 1991, p. 409.
K. H. Sun, *J. Am. Ceram. Soc.*, 1947, **30**, 277.
T. Takamori, in 'Treatise on Materials Science and Technology, vol. 17: Glass II', ed. M. Tomozawa and R. H. Doremus, Academic Press, New York, 1979, ch. 5.
M. Tomozawa, in 'Treatise on Materials Science and Technology, vol. 17: Glass II', ed. M. Tomozawa and R. H. Doremus, Academic Press, New York, 1979, ch. 3.
A. P. Tomsia, J. A. Pask, and R. E. Loehman, in 'Engineered Materials

Handbook, vol. 4: Ceramics and Glasses', ed. S. J. Schneider, Jr., ASM International, Materials Park, OH, 1991, p. 493.
A. Q. Tool, *J. Am. Ceram. Soc.*, 1946, **29**, 240.
F. V. Tooley, in 'The Handbook of Glass Manufacture', ed. F. V. Tooley, Books for Industry, New York, 1974, vol. 1, sect. 2.
F. V. Tooley, in 'Engineered Materials Handbook, vol. 4: Ceramics and Glasses', ed. S. J. Schneider, Jr., ASM International, Materials Park, OH, 1991, p. 402.
F. V. Tooley, M. A. Knight, A. K. Lyle, and V. C. Swicker, in 'The Handbook of Glass Manufacture', ed. F. V. Tooley, Books for Industry, New York, 1974, vol. 1, sect. 9.
D. R. Uhlmann, in 'Nucleation and Crystallization in Glasses', ed. J. H. Simmons, D. R. Uhlmann, and G. H. Beall, American Ceramic Society, Columbus, OH, 1982, p. 1.
D. R. Uhlmann, *J. Am. Ceram. Soc.*, 1983, **66**, 95.
A. K. Varshneya, 'Fundamentals of Inorganic Glasses', Academic Press, San Diego, CA, 1994, ch. 1, 4, 5, 7–10, and 18–20.
W. Vogel, 'Chemistry of Glass', American Ceramic Society, Columbus, OH, 1985, ch. 7, 8, and 11.
W. Vogel, in 'Optical Properties of Glass', ed. D. R. Uhlmann and N. J. Kreidl, American Ceramic Society, Westerville, OH, 1991, ch. 1.
M. C. Weinberg, *J. Non-Cryst. Solids*, 1994, **167**, 81.
J. E. White and D. E. Day, in 'Rare Elements in Glasses', ed. J. E. Shelby, Trans Tech Publications, Aedermannsdorf, Switzerland, 1994, p. 181.
F. E. Wooley, in 'Engineered Materials Handbook, vol. 4: Ceramics and Glasses', ed. S. J. Schneider, Jr., ASM International, Materials Park, OH, 1991, pp. 386–393.
B. Wunderlich, 'Thermal Analysis', Academic Press, San Diego, CA, 1990, ch. 6.
W. H. Zachariasen, *J. Am. Chem. Soc.*, 1932, **54**, 3841.

Subject Index